The Ultimate Mad Scientist Handbook

Other Books by Joey Green

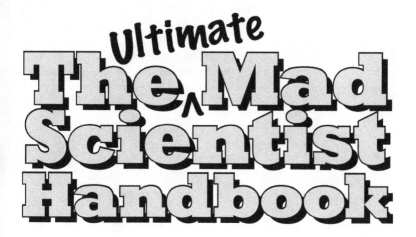

The *Ultimate* Mad Scientist Handbook

How to Make Your Own Balloon Rocket, Green Slime, Potato Popper, Edible Glass, Quicksand, Lava Lamp, and Much More!

Joey Green

Lunatic Press
Los Angeles

Material in this book was previously published in
The Mad Scientist Handbook (© 2000 by Joey Green),
The Mad Scientist Handbook 2 (© 2002 by Joey Green),
and *Potato Radio, Dizzy Dice* (© 2004 by Joey Green),
and has been revised, updated, and expanded.

Published by Lunatic Press, Los Angeles
www.lunaticpress.com

Book design by Joey Green
PRINTED IN THE UNITED STATES OF AMERICA

Library of Congress Control Number: 2012910454

ISBN: 0-0772590-6-4
ISBN-13: 978-0-9772590-6-9
10 9 8 7 6 5 4 3 2 1

For
Ashley, Julia,
Eric, Rebecca, Andrew, Lauren,
Matthew, Sammy,
Jonathan, Alexander, Zachary
Corinne, Janelle,
Kailey, Shannon, and Devin

Contents

Introduction

When I was in the fourth grade, I nearly blew up our house. My father helped me build a model volcano with metal screening in the shape of a cone tacked to a piece of wood, strips of newspaper, plaster of Paris, and brown paint. He bought a jar of strange orange powder from a chemical supply store. I poured the powder into the upside-down metal cap from an aerosol spray can built  into the top of the volcano, lit it with a match, and *KA-BOOM!*

While I didn't learn all that much about real volcanoes, I did learn that my dad was one very cool guy, although he never told me where on earth he bought that strange orange powder or what it was called—until now.

This book is for anyone who loves fun, freaky science experiments. It's a simple one-stop cookbook for making fake blood, green slime, edible glass, quicksand, underwater fireworks, and a huge mess in your kitchen and garage—all with stuff you've already got around the house (or things you can easily get). I've tested everything in this book. It all works. Guaranteed.

Plus I've provided a lot of weird and wild scientific facts, like why every experiment works and who thought it up, in case you just have to know. If you happen to learn something along the way (oh, like chemistry or physics) and wind up getting a Ph.D. at MIT in some quirky subject like rocket science, don't blame me. All I'm really interested in teaching you is how to make really cool slimy goo, strange doohickeys, and one heck of a mess.

—JOEY GREEN

Alka-Seltzer Rocket

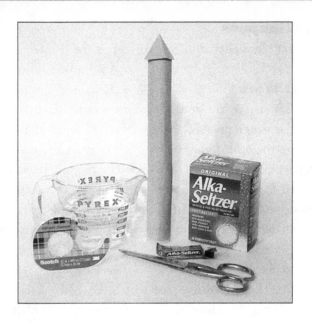

What You Need

- Colored construction paper
- Fuji 35mm film canister
- Scotch Tape
- Scissors
- Safety goggles
- 1 cup cold water
- Alka-Seltzer tablets

What To Do

Roll the sheet of construction paper around the film canister, so the open end of the film canister stick outs. Tape the paper onto the film canister. Using scissors and Scotch Tape, add a paper cone to the opposite end of the paper tube.

Wearing safety goggles, turn the rocket upside down so the open end of the film canister faces upward. Fill the canister half-

1

way with cold tap water. Drop in two Alka-Seltzer tablets. Snap on the lid, turn the rocket right-side up, set it down on the ground, and quickly step back.

What Happens
The rocket blasts approximately six feet off the ground.

Why It Works
On Fuji 35mm film canisters, the lid snaps inside the body. When activated in water, the Alka-Seltzer tablet releases carbon dioxide gas, filling the sealed canister—until the pressure becomes too great, popping the canister off its lid. The pressure blows the gas out of the canister, pushing the canister with an equal force in the opposite direction. As Sir Isaac Newton's third law of motion states: For every action there is an equal and opposite reaction.

Bizarre Facts
■ In 1928, Hub Beardsley, president of Dr. Miles Laboratories, discovered that the editor of the local newspaper in Elkhart, Indiana, prevented his staff from getting influenza during a severe flu epidemic by giving them a novel combination of aspirin and baking soda. Beardsley immediately set his chief chemist, Maurice Treneer, to work devising a tablet containing the two ingredients.
■ The "Plop, Plop, Fizz, Fizz, Oh What a Relief It Is!" vintage theme song for Alka-Seltzer, written by Tom Dawes in 1977, remains one of the most recognized commercial melodies and a favorite of popular culture trivia buffs.
■ The only man-made structure visible from space is the Great Wall of China.
■ If the Spaceship Earth geosphere at the Epcot theme park at Walt Disney World were a golf ball, to be the proportional size to hit it, you would be more than one mile tall.

Anti-Gravity Bucket

What You Need

- ❑ Plastic bucket with handle
- ❑ Clothesline rope (6 feet in length)
- ❑ 2 quarts water
- ❑ Work gloves

What To Do

Tie one end of the rope to the handle of the bucket. Pour the water into the bucket. Wearing the gloves (to avoid getting rope burn), wrap the free end of the rope around one hand. Standing outdoors, swing the bucket from the rope like a lasso, making the bucket circle around you. Continue swinging the bucket around so it gathers enough momentum to circle around you parallel to the ground. Then tip the angle of the rope so the bucket circles at an angle over your head.

What Happens

The water remains inside the circling bucket.

Why It Works

Centrifugal force—the force created by whirling the rope—causes the water to move toward the bottom of the bucket. Centripetal force, however, holds the water in the bucket. The rope applies centripetal force to the whirling bucket, pulling it inward and preventing the bucket (and the water) from moving in a straight line.

Bizarre Facts

■ The faster your twirl the bucket around you, the stronger the pull (or centrifugal force) on the rope.

■ Most people confuse centrifugal force with centripetal force. Centrifugal force is the outward force on the object rotating about an axis. Centripetal force is the inward force on that same object. The forces acting on the same object are equal and opposite, in keeping with Newton's third law of motion (For every action there is an equal and opposite reaction).

■ When you ride on a merry-go-round, your body wants to shoot off in a straight line, but is held back by centripetal force (your grip on the horse).

■ The earth's gravity exerts a centripetal force on the moon and artificial satellites, preventing them from shooting out into space and keeping them orbiting the planet. Similarly, the sun's gravity exerts a centripetal force on the earth, preventing our planet from shooting out into space.

■ A spinning object, such as a top, stays in motion due to angular momentum.

■ To perform a spectacular spin, an ice skater starts spinning with his arms outstretched, creating a larger spin diameter and greater angular momentum. By suddenly pulling in his arms, the skater reduces his diameter, causing his body to spin faster to conserve angular momentum.

Anti-Gravity Machine

What You Need

❑ Three books, each at least 1-inch thick
❑ Two yardsticks
❑ Two plastic funnels of equal size
❑ Black electrical tape

What to Do

Stack two books on top of each other on the floor. Place the third book far enough away on the floor so you can lay a yardstick across the books to form a bridge.

Place the second yardstick next to the first yardstick to form a V-shape with the open end of the V on the stack of two books. Tape the bowls of the funnels together. Place the joined funnels on the lower end of the track formed by the yardsticks.

What Happens

The joined funnels roll up the incline.

5

The Fall of Aristotle

The Greek philosopher and scientist Aristotle claimed that heavy objects fall faster than light objects, a widely accepted belief until the sixteenth century when, according to legend, Italian scientist Galileo Galilei simultaneously dropped two iron balls from the Leaning Tower of Pisa (proving that all objects fall at the same rate of acceleration).

Why It Works

Although the joined funnels appear to defy the laws of gravity, in reality, their center of gravity (the point at which the effect of gravity on an object seems to be concentrated) moves downward as the joined funnels move along the inclined yardsticks.

Bizarre Facts

■ The center of gravity of the hollow joined funnels is at its center, even though there is no matter at that point for gravity to affect.

■ When a boomerang is thrown, it spins about its center of gravity, which is outside its body, between the arms of the V.

■ The moon's gravitation causes the ocean tides on earth.

■ You would weigh 0.526 percent more if you were standing on the North Pole than you would if you were standing on the equator. An object on earth does not weigh the same at all places on the planet because the earth rotates and it is not perfectly round.

■ While Galileo Galilei is credited with determining that falling objects fall at the same rate, Giambattista Benedetti determined the exact same thing in 1553—eleven years before Galileo was born.

■ Sir Isaac Newton, considered a poor student in school, discovered gravity, invented calculus, and, in 1699, became master of the mint in England, prosecuting counterfeiters.

Balloon Car

What You Need
- ❏ Krazy Glue
- ❏ Small plastic spool
- ❏ Old compact disc
- ❏ Button
- ❏ Medium-sized balloon

What to Do
Using the Krazy Glue, adhere the plastic spool onto the compact disc so the hole in the spool is directly over the hole in the compact disc. Let dry. Glue a button over the top hole in the spool so the holes in the button are directly over the hole in the spool. Inflate the balloon, pinch the neck to prevent the air from escaping, and stretch the lip of the balloon over the spool. Set the compact disc on a flat tabletop and let go of the balloon.

What Happens
The compact disc floats across the table like a hovercraft.

Why It Works
An invisible cushion of air acts as a lubricant and reduces friction between the compact disc and the tabletop, the same way adding oil to a car engine prevents the parts from rubbing against each other.

7

Bizarre Facts

■ The word *balloon* is slang for "a hobo's bedroll."

■ A compact disc holds three miles of playing track and is read from the inside edge to the outside edge, the reverse of how a vinyl record works.

■ In 1954, English electronics engineer Christopher Cockerell designed the hovercraft by attaching two tin cans, one inside the other, to an industrial air blower mounted on a stand and blowing air through the gap between the tin cans.

■ The SRN-4 Mark III, the world's largest civil hovercraft, weighed 305 tons, carried up to 418 passengers and sixty cars across the English Channel from 1968 to 2000, and traveled at a top speed of 75 miles per hour—nearly twice the speed of the fastest ocean liner.

■ The Russian Zubr class LCAC, the world's largest military hovercraft, can transport three T-80 main battle tanks, ten armored vehicles with 140 fully-equipped troops, 500 troops, or up to 130 tons of cargo, and travels at a top speed of 46 miles per hour.

Know-It-All

In 1899, Charles H. Duell, Commissioner of the U.S. Office of Patents, claimed: "Everything that can be invented has been invented."

Balloon Rocket

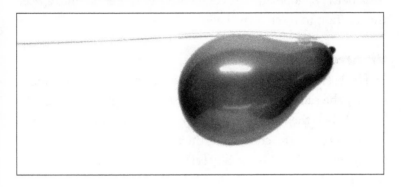

What You Need

- Piece of string, 10 to 25 feet in length
- Drinking straw
- Balloon
- Scotch Tape

What to Do

Tie one end of the string to a tree or post. Thread the straw onto the free end of the string, and then tie that end of the string to a second tree or post, making sure the string is taut. Move the straw to one end of the string.

Inflate the balloon and pinch the neck to prevent the air from escaping. Use two pieces of Scotch Tape to attach the inflated balloon to the straw so the balloon is parallel to the straw and the mouth of the balloon points toward the closest post or tree.

Release the balloon.

What Happens

The balloon and straw jet across the string until the balloon completely deflates.

Why It Works

As Sir Isaac Newton's third law of motion states: For every action there is an opposite and equal reaction. When you release the neck of the balloon and the compressed air rushes out into space, the reaction to it sends the balloon forward.

Bizarre Facts

■ The basic principle behind a balloon zooming across a string is exactly the same principle behind a space rocket launching into space. When the fuel burns, gas escapes from the rocket's bottom, and the opposite reaction sends the rocket upward.

■ The ocean liner *Queen Elizabeth II* moves approximately fifty feet for each gallon of diesel fuel it burns.

■ The longest recorded flight of a chicken is thirteen seconds.

■ In 1895, Lord Kelvin, president of the Royal Society, said: "Heavier-than-air flying machines are impossible."

■ In 1921, responding to rocket scientist Robert Goddard's revolutionary work, the *New York Times* editorialized: "Professor Goddard does not know the relation between action and reaction and the need to have something better than a vacuum against which to react. He seems to lack the basic knowledge ladled out daily in high schools." Five years later, Goddard launched the first liquid fuel rocket.

■ In 1927, Felix the Cat became the first cartoon character made into a balloon for the Macy's Thanksgiving Day Parade.

In Space No One Can Hear You Scribble

During the space race in the 1960s, NASA spent $1 million to develop a ballpoint pen that would write in zero gravity. The Soviet Union solved the same problem by giving their cosmonauts pencils.

Battery Madness

Money Battery

What You Need

- Quarter
- Coffee filters
- Pencil
- Scissors
- Twelve pennies (dated after 1983)
- Twelve ¼ zinc washers
- Wire cutters
- 2-foot length of 22-gauge electrical wire
- Nail
- Black electrical tape
- Bowl
- Lemon juice
- Compass

What To Do

Place the quarter on a coffee filter and trace around it with the pencil to create a circle. Repeat this eleven times, creating a total of twelve circles. With the scissors, cut out the twelve circles.

11

Place a penny on the table-top. Place a circle of coffee filter on top of the penny. Place a zinc washer on top of the coffee filter. Place another penny on top of the zinc washer, followed by a second circle of coffee filter, followed by another zinc washer. Repeat until

you have stacked all the pennies, circles of coffee filter, and zinc washers—ending with a zinc washer on top.

With adult supervision, use the wire cutters to strip one inch of plastic coating off each end of the wire. Wrap the middle of the wire around the nail tightly until the nail is covered with wire. Peel off a four-inch strip of electrical tape. With the scissors, carefully cut the strip of tape down the center to make two narrow four-inch strips of tape. Using one strip of the tape, attach the end of one wire to the top coin and the end of the second wire to the bottom coin, while simultaneously securing all the coins together (as if using a rubber band).

Fill the bowl with lemon juice. Gently drop in the stack of coins. Move the head of the nail over the point of the compass needle.

What Happens

The coffee filter disks, saturated with lemon juice and sandwiched

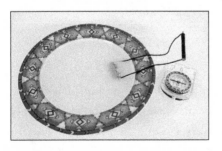

between two disks made from different metals, creates a wet cell battery with enough voltage to make an electro-magnet with the strength to move a compass needle.

Why It Works

An electrolyte (the citric acid in the lemon juice) between two electrodes of different chemically active materials causes one of them, called an anode (the copper in the pennies), to become negatively charged, and the other, called a cathode (the zinc in the washers), to become positively charged.

Pot Scrubber Battery

What You Need

❏ 2-foot length of 22-gauge electrical wire
❏ Wire cutters
❏ Nail
❏ Copper scouring pad

❏ Paper towels
❏ White vinegar
❏ Aluminum foil
❏ Compass

What To Do

With adult supervision, use the wire cutters to strip one inch from both ends of the wire. Wrap the middle of the wire around the nail tightly until the nail is covered with wire.

Secure one end of the wire to the cooper scouring pad. Saturate a sheet of paper towel with vinegar and then wrap it tightly around the copper pad. Wrap a sheet of aluminum foil tightly around the paper towel. Attach the other end of the wire to the aluminum foil.

Move the head of the nail over the point of the compass needle.

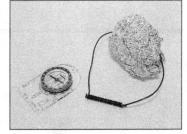

What Happens

The compass needle moves.

Why It Works

The vinegar (containing acetic acid) is the electrolyte, the cooper scouring pad is the anode, and the aluminum foil is the cathode, creating a wet cell battery with enough voltage to make an electromagnet with the strength to move a compass needle.

Bizarre Facts

■ In the 1790s, Italian scientist Count Alessandro Volta made the first battery by stacking pairs of silver and zinc disks separated from one another by cardboard disks moistened with a salt solution. The volt, a unit of electric measurement, is named after him.

■ Most pennies minted before 1983 are made from a copper-zinc alloy, while pennies minted after 1983 are made from zinc coated with a thin layer of copper.

As Primitive As Can Be

In an episode of *Gilligan's Island* [Episode 18: "X Marks the Spot"], to recharge the batteries in the castaways' radio, the Professor sets up several metal strips, pennies, and coconut shells filled with seawater.

Beach Ball Elevator

What You Need

- 4-foot length of vinyl flexible hose (¼ inch in diameter)
- Beach ball (1 foot in diameter)
- Black electrical tape
- Scissors
- Basketball hand pump
- Plank of ¾-inch thick pinewood (1-by-5 feet)

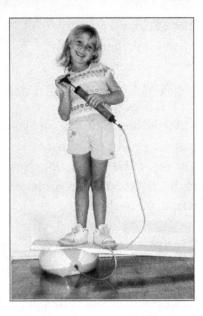

What To Do

Insert one end of the flexible hose into the open nozzle of the uninflated beach ball and secure in place with a piece of the electrical tape. Insert the nozzle of the basketball hand pump into the other end of the flexible hose and secure in place with another piece of electrical tape. Set the plank of wood on a clean, firm surface (sidewalk, tile floor) and place the uninflated beach ball under one end of the plank of wood. Stand on the other end of the plank to hold it in place. Have someone stand on top of the wood over the beach ball. Pump the basketball pump.

What Happens

As the beach ball fills with air, the plank of wood slowly rises, elevating the person standing on it.

Why It Works

French scientist and philosopher Blaise Pascal (1632–1662) discovered that a fluid in a container transmits pressure equally in all directions. This principle, known as Pascal's Law, explains how

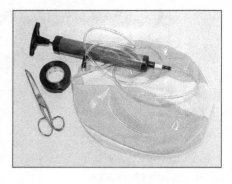

hydraulic lifts work. The air blown into the beach ball distributes pressure equally throughout the ball. If air at ten pounds of pressure per square inch is blown into the ball and ten square inches of the ball touch the piece of wood, the air will lift one hundred pounds of weight.

Bizarre Facts

■ In 1993, Amtrak's X2000 high-speed train began operating on the Washington–New York Metroliner route, capable of traveling up to 155 miles per hour and using a hydraulic tilting system to take curves 40 percent faster.

■ Hydraulic lifts allow specially designed city buses and vans to rise and lower to the curb to accommodate physically challenged passengers and those confined to wheelchairs.

■ The Bastille Opera House in Paris, France, contains a modular concert hall whose height, seating, and proscenium opening can all be altered with hydraulic lifts.

Water Under the Bridge

The world *hydraulic* stems from the root *hydro*, meaning "water"—despite that fact that hydraulic machines usually use liquids and gases other than water in their workings.

Blacklight Ink

What You Need

- Liquid Tide
- Paper cup
- Paintbrush
- Black posterboard
- Blacklight

What To Do

Fill a paper cup half way with Liquid Tide. With the paintbrush, paint designs or a message on the black posterboard. Let sit until the liquid detergent dries, turning invisible. In a dark room, turn on the blacklight and hold the poster near the light.

What Happens

The detergent painted on the black posterboard appears invisible in ordinary light, but glows purplish-white under the blacklight.

Why It Works

Liquid Tide contains a fluorescent chemical that is activated by the ultraviolet rays produced by a black light. The fluorescent

17

chemical in the Liquid Tide converts the ultraviolet light into visible light.

Bizarre Facts

■ Murine Tears eye drops glow under a blacklight as fluorescent yellow.

■ Cat urine glows under a blacklight.

■ Ultraviolet rays (also known as black light) are a form of light invisible to the human eye and lie beyond the violet end of the visible spectrum.

■ *Consumer Reports* claims, "No laundry detergent will completely remove all common stains," and reports very little difference in performance between major name-brand powdered detergents.

■ Ultraviolet rays can cause sunburn and penetrate clouds, which is why a person can get sunburned on an overcast day.

■ Lightning produces ultraviolet rays.

■ A blacklight lamp produces ultraviolet rays when an electric current passes through the glass tube filled with a gas or vapor.

■ Ultraviolet rays with wavelengths shorter than 320 nanometers can be used to kill bacteria and viruses.

■ Ultraviolet rays from the sun produce vitamin D in the human body.

Blacklight Bubbles

For a totally psychedelic experience, mix ½ cup Tide with ½ cup water, and use the solution to blow bubbles under the blacklight. The bubbles will glow purplish white.

Bottle Rocket

What You Need

- Safety goggles
- Electric drill with $\frac{1}{32}$-inch bit
- Cork
- Scissors
- Playtex Living glove
- Needle adapter for inflating balls
- Ruler
- Pencil
- Foam-core board
- X-acto knife or single-edge razor blade
- Two clean, empty 1-liter plastic soda bottles
- Clear packaging tape
- 1 cup water
- Funnel
- Bicycle pump

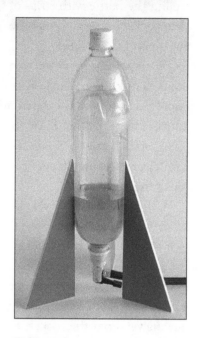

What To Do

With adult supervision and wearing safety goggles, drill a hole through the center of the cork. Using scissors, cut a section from a finger of the glove to fit tightly around the cork as a sleeve. Insert the needle adapter through the hole in the cork so it fits snugly.

On the foam-core board, measure a right triangle with a base 3 inches long and a height 8 inches long. With adult supervision, carefully cut out the triangle with the X-acto knife or single edge razor blade. Use this triangle as a template to create two more triangles.

19

Turn one of the plastic bottles upside down and attach the triangular fins at equal distances around the bottle with strips of clear packaging tape, so that the fins support the rocket.

With adult supervision, use the X-acto knife or single-edge razor blade to cut off the top from the second plastic soda bottle—4 inches from the top. Place the top over the bottom of the first plastic soda bottle to create the rocket cone and secure in place with clear packaging tape. Secure the bottle cap in place on the top of the rocket.

Using the funnel, hold the rocket upside down and fill the bottle with 1 cup water. Insert the cork into the neck of the bottle so it fits tightly. Connect the bicycle pump to the needle adapter.

Set the rocket on its fins on a flat surface outdoors—away from trees, telephone wires, electrical cables, and buildings. With adult supervision, wearing safety goggles, and standing a safe distance away from the rocket, pump air into the bottle using the bicycle pump.

What Happens

The cork eventually pops from the neck of the bottle, and water and air shoot out, sending the rocket high into the air.

Why It Works

The air pressure builds up inside the bottle until the cork pops out, allowing the water and air to escape from the bottle. Consequently, the bottle rocket is propelled upward because, as Sir Isaac Newton stated in his third law of motion, for every action there is an opposite and equal reaction.

Bizarre Facts

■ The Space Shuttle and *Saturn V* rockets are propelled into space due to Newton's third law of motion. When fuel burns inside the rocket's combustion chamber, the resulting hot gases shoot from the nozzle. The reacting force sends the rocket upwards.

■ Sir Isaac Newton, an ordained priest in the Church of England, taught mathematics at Cambridge University.

■ Newton's third law of motion explains why guns kick back violently when fired.

■ When you shoot a cue ball into a stationary ball on a pool table, both balls push each other with equal force in opposite directions. The stationary ball gains momentum and the cue ball loses the exact same amount of momentum.

Making a Splash

You can demonstrate Newton's third law by simply jumping off an inflatable raft in a swimming pool. When you jump forward, the raft moves backward with equal momentum.

Brainy Ball

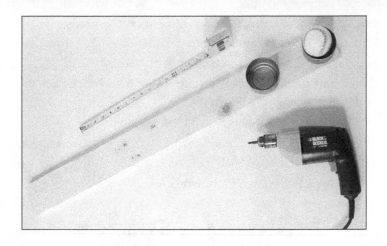

What You Need

- Electric drill with a Phillips screwdriver bit
- Two ½-inch wood screws
- Two clean, empty 6-ounce tuna fish cans
- Plank of ¾-inch-thick pinewood (3-by-36 inches)
- Yard stick or tape measure
- Baseball

What To Do

With adult supervision, use the drill to screw one tuna fish can at the end of the piece of lumber. Screw the second can to the lumber, three inches away from the first can.

Place the piece of lumber on the floor, perpendicular to a wall, and slide the free end against the wall. Place the baseball in the can nearest the end, lift that end of the lumber thirty inches from the ground (leaving the other end on the ground against the wall). Let the piece of wood drop to the ground.

What Happens

When the lumber falls to the floor, the baseball in the first can falls into the second can.

Why It Works

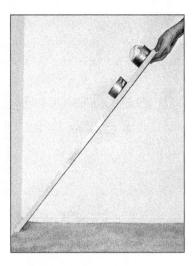

Since gravity pulls all falling objects to earth at the same speed, you would think the baseball would stay in the first can. However, the piece of wood is not a free-falling object. The end touching the floor does not fall at all, and it actually pulls down the rest of the wood faster than gravity. The baseball, unattached to the wood or can, falls freely at the constant rate of 32 feet per second. The can drops faster than the baseball, the baseball falls straight down, the can (attached to the wood) follows the arc of a circle to land below the falling baseball.

Bizarre Facts

■ The piece of wood's center of gravity falls at the speed of gravity. When one end of the piece of wood is lifted up, the lumber's center of gravity is approximately twelve inches away from the free end.

■ If you run eastward (in the direction of the earth's spin) you weigh fractionally less than you do standing still. The mild centrifugal force created by the earth's rotation counteracts the effect of gravity. If you run westward (against the earth's spin), you will weigh slightly more. A 40,000-ton ship sailing east at 20 knots along the equator (where the centrifugal effect is greatest) weighs approximately six tons less than the same ship sailing west.

■ You weigh less standing on the top of a mountain than you do standing at sea level (because the force of gravity diminishes the farther you are from the center of the earth).

The Apple Doesn't Fall Far from the Tree

Most of Sir Isaac Newton's early biographers fail to mention the story that Newton discovered gravity after watching an apple fall from a tree—casting serious doubt over whether the incident ever really occurred. The two sources of the tale are the French philosopher Voltaire and the Reverend William Stukeley, neither of whom witnessed the actual event.

Voltaire, in his book *Elements of Newtonian Philosophy*, published in 1738 (eleven years after Newton's death and seventy years after the alleged incident), reported that Newton told the story to his niece, Catherine Barton Conduitt, who cared for Newton in his later years.

Reverend Stukeley, in his biography of Newton (written in 1752, but not published until 1936), reported that the physicist told him about the incident while the two were having tea together in the apple orchard at Newton's home. Most scholars believe Newton embellished the story.

Butter Machine

What You Need

❏ 1 pint of whipping cream
❏ Clean, empty, resealable airtight plastic container
❏ Yellow food coloring
❏ Spoon

What To Do

Pour the entire pint of cream into the plastic container, seal the lid, and shake the sealed container for fifteen minutes or more. Slowly pour out the milky liquid, leaving the butter in the container. Add two drops of yellow food coloring and mix the butter with a spoon. Store the butter in the refrigerator.

What Happened

The cream separates into butter and a milky liquid called buttermilk.

Why It Works

Milk and cream contain an emulsion of butterfat in the form of tiny, invisible droplets. Shaking or churning causes the butter granules to cling together to form butter.

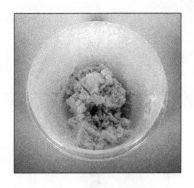

Bizarre Facts

▨ Ancient Romans used butter as a hairdressing cream and as a skin cream.

▨ Cream contains roughly ten times more butterfat that milk does.

▨ Cream can be made by simply allowing a glass of raw milk to stand undisturbed overnight. Gravity causes the milk to sink to the bottom of the glass and the cream to rise to the surface. The layer of cream can be spooned off. The remaining milk is called skimmed milk.

▨ Wisconsin produces more butter than any other state, followed by California and Minnesota.

▨ The natural color of butter varies from pale to deep yellow, depending on the breed of cow and what it was fed. Cows eating fresh green grass produce a deep yellow butter while cows eating grain or hay produce a paler color butter. Butter makers usually add food coloring to butter to make it more attractive to consumers.

▨ Machines cut butter into rectangular sticks called *prints*.

▨ Butter can be made from the milk of cows, goats, horses, reindeer, sheep, yaks, and other animals.

Buffalo Butter

People started making butter as early as 2000 B.C.E., when people in India began making butter from the milk of water buffaloes.

Confetti Blaster

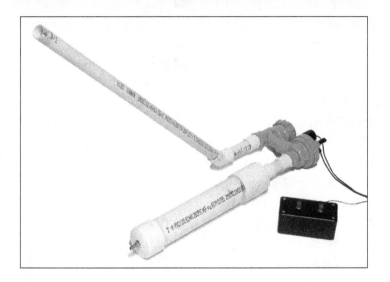

What You Need

- Safety goggles
- Electric drill with ⅜-inch bit and ¼-inch bit
- 2-inch-diameter PVC end cap
- Metal automotive tire valve with nuts to tighten (available at an automotive supply store)
- Adjustable wrench
- Utility knife
- Teflon tape
- 24-volt plastic sprinkler valve with ¾-inch diameter threaded male pipe connections
- Two PVC pipe adapters from ¾-inch to ¾-inch female screw head
- 2-foot length of 2-inch-diameter PVC pipe
- 2-inch-diameter PVC coupler
- PVC pipe reducer fitting from 2-inch diameter to ¾-inch diameter

27

- 3-inch length of ¾-inch-diameter PVC pipe
- 3-foot length of ¾-inch-diameter PVC pipe
- ¾-inch 45-degree-angle PVC coupler
- 6-inch length of ¾-inch-diameter PVC pipe
- PVC pipe primer
- PVC pipe glue
- Wire cutters
- 15-foot length of two-conductor 24-gauge wire
- Black electrical tape
- Plastic project encloser box (5-by-2½-by-2 inches)
- SPST on/off toggle switch
- SPST momentary push-button switch
- Two 9-volt battery snap connectors
- 6-inch length of single conductor 24-gauge wire
- Solder gun
- Solder
- Two 9-volt batteries
- Phillips screwdriver
- Bicycle pump with air pressure gauge
- Tissue
- 4-foot length of wooden dowel (¼-inch diameter)
- Box of foam peanuts
- Wax paper
- Masking tape

What To Do

With adult supervision and wearing safety goggles, drill a ⅜-inch hole in the center of the end cap. Unscrew the tightening nut from the metal automotive tire valve, insert the tire valve through the inside of the end cap (making sure the rubber gasket covers the drilled hole), and tighten the nut securely in place with the adjustable wrench.

Using the utility knife, carefully scrape off any burrs left from the ends of the cut pieces of PVC pipe to assure a clean seal when the pipes are securely sealed together airtight.

Wrap a piece of Teflon tape around each one of the threaded male joints on the 24-volt sprinkler valve. (Do not use the PVC pipe cement on any threaded head.) Screw a PVC pipe adapter (with the threaded female joint) onto each one of the threaded male joints of the sprinkler valve. Secure tightly.

Coat the outsides of the ends of each PVC pipe and the inside of each PVC coupler, the outside and the inside of the PVC pipe reducer fitting, and the end cap with PVC pipe primer. Using the PVC pipe glue, attach the 2-inch-diameter PVC end cap (previously prepared with the tire valve) to the end of the 2-foot length of 2-inch-diameter PVC pipe. To the other end of the 2-foot length of 2-inch-diameter PVC pipe, glue the 2-inch-diameter PVC coupler. Glue the PVC pipe reducer fitting inside the 2-inch-diameter PVC coupler. Glue one end of the 3-inch length of ¾-inch-diameter PVC pipe into the ¾-inch hole of the PVC pipe reducer fitting. Glue the open end of the 3-inch length of PVC pipe into the adapter at the bottom of the 24-volt sprinkler valve beneath the chamber with the wires (and the top secured with eight screws). This completes the air-compression chamber.

Glue one end of the 3-foot length of ¾-inch-diameter PVC pipe into the ¾-inch 45-degree-angle PVC coupler. Glue one end of the 6-inch length of ¾-inch-diameter PVC pipe into the open end of the 45-degree-angle PVC coupler. Glue the open end of the 6-inch length of ¾-inch-diameter PVC pipe into the adapter at the bottom of the 24-volt sprinkler valve beneath the chamber without the wires. Make certain that when the device sits on the floor, the 3-foot length of pipe rises from the floor at a 45-degree angle. This completes the barrel of the confetti blaster. Allow the PVC pipe glue to dry for 24 hours before use.

Use the wire cutters to strip ½ inch of plastic coating off the four ends of the two-conductor wire. Wire one end of each wire to the ends of the wire coming from the 24-volt sprinkler valve. Individually wrap each connection with electrical tape to prevent the connections from contacting each other. Then tape the two wrapped connections together.

Drill two ¼-inch holes 2 inches apart in the top of the plastic project encloser box. Insert the on/off toggle switch into one hole and secure with the nut. Insert the momentary push-button switch into the other hole and secure with the nut.

Drill a hole in the center of one of the 5-inch-long sides of the plastic box itself. Tie a knot in the free end of the 15-foot wire, 6 inches from the two free ends. Insert the two free ends into the hole you drilled in the side of the plastic box and tie another knot in the double wire inside the box to secure the wire in place.

Wire one of the free wires to one wire from a 9-volt battery snap connector. Wire the free wire from the 9-volt battery snap connector to one of the wires from the second 9-volt battery snap connector. Wire the remaining free wire from the second 9-volt battery snap connector to one pole on the on/off toggle switch.

Use the wire cutters to strip ½ inch of plastic coating off the two ends of the 6-inch length of single-conductor wire. Wire one end of the 6-inch-long wire to the free pole on the on/off toggle switch. Wire the remaining end of the 6-inch length of wire to a pole on the momentary push-button switch. Wire the remaining free end of the 15-foot wire to the free pole on the momentary push-button switch.

Solder the connections. When the solder cools, wrap the connections with electrical tape (to prevent the batteries that will be inside the box from accidentally bridging the contacts). Snap the batteries into place. (The two 9-volt batteries provide a total of 18 volts, which, surprisingly, is sufficient to activate the 24-volt sprinkler valve.) Using a Phillips screwdriver, screw the cover panel into place on the project enclosure box.

After allowing the PVC pipe glue to dry for 24 hours before using the blaster, attach the bicycle pump to the tire valve. Pump air into the chamber until you achieve 80 pounds per square inch (psi). Fill a bathtub with 6 inches of water, and hold the device underwater to check for leaks. If you encounter a leak, use an aerosol can of Fix-a-Flat (available at an automotive store) to fix the leak.

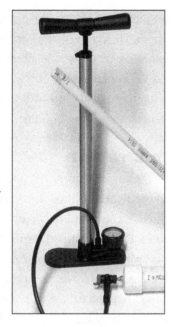

Ball up a tissue, insert it into the open end of the ¾-inch pipe, and use the dowel to push the tissue down the tube just before the first joint. Break the foam peanuts into small pieces, and pour one cup of them into the open end of the ¾-inch pipe.

Cover the open end of the barrel with a piece of wax paper and secure it in place with masking tape around the circumference (not over the top of the wax paper).

Wearing safety goggles, attach the bicycle pump to the tire valve and pump air into the chamber until you achieve 80 psi. (A 24-volt sprinkler valve can generally handle up to 150 psi.)

Position the confetti blaster on the ground outside, aiming the open end of the ¾-inch pipe away from people or pets and toward an open area. Flip the on-off switch to on. Press the momentary push-button switch.

What Happens

The foam peanuts blast through the wax paper, causing a popping sound, and shoot up to forty feet into the air.

Why It Works

When you press the button, the batteries activate the electromagnet inside the sprinkler valve, opening the value and allowing the air pressure stored in the first chamber to rush out the barrel of the air gun, forcing out the contents of the barrel. The homemade air cannon can also shoot off streamers or confetti. To obtain streamers or confetti, visit www.artistryinmotion.com. You will need confetti that fits into a ¾-inch cannon. To shoot larger loads of confetti, use 1-inch-diameter PVC pipe for the cannon.

Bizarre Facts

▨ Professional confetti cannons commonly use canisters of compressed carbon dioxide to power the blasts. Canisters of carbon dioxide can be quite costly.

▨ In 1839, German pharmacist Eduard Simon isolated a strong, versatile substance from natural resin, without knowing what he had discovered. Nearly a century later, German chemist Hermann Staudinger identified Simon's discovery, comprised of long chains of styrene molecules, as a plastic polymer—polystyrene.

▨ In the 1940s, Dow Chemical scientist Ray McIntire, working to develop a new flexible electrical insulator, tried to create a new rubber-like polymer by combining styrene with isobutylene, a volatile liquid, under pressure. Instead of producing an elastic polymer, his experiment accidentally yielded foam polystyrene, filled with air bubbles and thirty times lighter than regular polystyrene.

▨ In 1954, the Dow Chemical Company introduced its polystyrene foam product under the brand name Styrofoam. The company first introduced Styrofoam as a flotation material in life rafts and lifeboats and later as a thermal insulation material.

▨ Polystyrene foam products are 95 percent air and only 5 percent polystyrene.

▨ Polystyrene foam products are produced primarily using two types of blowing agents: pentane and carbon dioxide.

Copper Nail

What You Need

- ¼ cup white vinegar
- ⅛ teaspoon salt
- Clean, empty glass jar
- Iron nail
- Baking soda
- Sponge
- Twenty pennies (dated after 1983)

What to Do

Pour the vinegar and salt into the jar. Stir well. Clean the nail with the baking soda and sponge. Rinse thoroughly. Put the pennies into the vinegar-and-salt solution for three minutes. Drop the clean nail into the vinegar-and-salt solution with the pennies. Wait fifteen minutes.

What Happens

The nail is coated with copper, and the pennies are shiny clean.

Why It Works

The acetate in the vinegar (also known as acetic acid) combines with the copper on the pennies to form copper acetate, which then adheres to the iron nail.

Bizarre Facts

Heads, You Lose

A penny will be tossed heads 49.5 percent of the time. The head side weighs 0.5 percent more than the tail side, so it tends to land downward.

■ The word vinegar is derived from the French words *vin* (wine) and *aigre* (sour).

■ The oldest way to make vinegar is to leave wine made from fruit juice in an open container, allowing microorganisms in the air to convert the ethyl alcohol to acetic acid.

■ Vinegar lasts indefinitely in the pantry without refrigeration.

■ Hannibal, the Carthaginian general, used vinegar to help clear boulders blocking the path of his elephants across the Alps. Titus Livius reported in *The History of Rome* that Hannibal's soldiers heated the rocks and applied vinegar to split them.

■ According to the *New Testament*, Roman soldiers offered a sponge filled with vinegar to Jesus on the cross. While the act is considered cruel, vinegar actually shuts off the taste buds, temporarily quenching thirst, suggesting that the Roman soldiers may have been acting out of kindness.

■ Copper is the best low-cost conductor of electricity.

■ Copper does not rust. In damp air, copper gets coated with a green film called a *patina*, protecting it against further corrosion.

■ Pennies minted after 1983 are actually made from zinc coated with 2.5 percent copper.

Cosmic Ray Detector

What You Need

- Scissors
- Black heavy felt
- Clean, empty glass jar with a rubber washer or cardboard filler under the lid
- Ruler
- Rubber cement
- Black velvet
- Rubbing alcohol
- Two towels
- Pie tin
- Work gloves
- Block of dry ice (enough to fill the pie tin)
- Hammer
- Ice tongs
- Flashlight
- Three books
- Masking tape
- Small magnet

What to Do

Cut the felt in a circle to lie in the bottom of the jar. Glue the felt in place with rubber cement. Cut a strip of felt 1 inch wide and glue it around the inside wall at the bottom of the jar.

Cut a strip of velvet 1 inch wide and glue it to the top of the jar's inside wall. Cut a circular piece of velvet to fit inside the metal lid, over the rubber or cardboard, and glue it in place.

Pour enough rubbing alcohol into the jar to saturate the felt at the bottom of the jar thoroughly and cover the bottom of the jar. Screw the lid on the jar tightly and let it sit for ten minutes.

Spread a towel on a flat surface and place the pie tin on top of it. Wearing work gloves (to avoid touching the dry ice with your bare hands), wrap the second towel around the block of dry ice, and then break up the block with the hammer. Using the ice tongs, put the cubes of dry ice into the pie tin, making a level surface on which to place the jar.

Turn the jar upside down and place it on the dry ice in the center of the pie tin.

Position the flashlight on top of the books, attaching it in place with masking tape, aiming the beam through the lower half of the jar. Turn off the lights, and cover the bottom of the jar with the palm of your hand to warm the alcohol-soaked velvet. Observe carefully.

What Happens

Within five minutes, the alcohol vapor condenses, warmed by your hand, and you'll see a continuous rain of fine mist 1 inch below the top of the chamber. After another five minutes, the rain decreases.

When the interior temperature is just right, you see cobweb-like threads suddenly appearing and disappearing at various angles about an inch above the lid. These are vapor trails made by cosmic rays passing through the jar. Place the magnet against the side of the jar, and the trails will be deflected toward it.

Why It Works

The heat from your hand warms the top of the chamber, and the dry ice cools the bottom of the chamber. Somewhere between these temperature extremes, usually about 1 to 2 inches from the bottom, the air becomes saturated with alcohol vapor, and particle trails become visible where cosmic rays cause the alcohol to condense.

Bizarre Facts

■ Cosmic rays are high-energy particles that originate from explosions in outer space—from phenomena such as supernovas and pulsars. At this very moment, cosmic rays are hurtling through space at nearly the speed of light.

■ An average of three to six cosmic ray particles strike each square inch of the earth's atmosphere every second.

■ Cosmic rays are penetrating your body at this very moment.

■ Physicist Murray Gell-Mann named the subatomic particles known as quarks after a line from James Joyce's novel *Finnegans Wake*: "Three quarks for Muster Mark!"

■ The common goldfish can see both infrared and ultraviolet light.

The Lies Dictators Tell

Nazi dictator Adolf Hitler, refusing to accept the fact that a Jew had come up with the theory of relativity, falsely claimed that Albert Einstein had stolen the idea from some papers found on the body of a German officer who had been killed in World War I.

Dancing Ping-Pong Ball

What You Need

❑ 5-inch-long plastic comb
❑ Wool sweater
❑ Ping-Pong ball

What To Do

On a cold day, rub the comb on the wool sweater for one minute.
Place the Ping-Pong ball on a smooth, flat surface. Hold the comb
vertically an inch above the Ping-Pong ball and slowly spin the
comb in circles above the ball.

What Happens

The Ping-Pong ball dances and spins, following the movements
of the comb.

How It Works

Like charges repel, unlike charges attract. Rubbing the plastic comb on the wool sweater charges the comb with static electricity—transferring electrons from the wool to the comb and giving the comb a negative charge. The Ping-Pong ball remains uncharged. These unlike charges attract each other, drawing the uncharged Ping-Pong ball towards the charged comb.

Bizarre Facts

■ If you touch the negatively charged comb to the Ping-Pong ball, the ball will become negatively charged and be attracted to the comb.

■ When you comb your hair quickly on a dry day, your hair loses electrons and becomes positively charged, while the comb gains electrons and becomes negatively charged. As you comb your hair, the static electricity generated makes it crackle.

■ When you walk across carpet on a cold day, you generate static electricity. If you touch a metal object, like a door knob, the positive charge you have generated will leap to the uncharged door knob—creating a spark and giving you a slight shock.

■ The earliest known comb is believed to be the dried backbone of a large fish.

■ Man-made combs dating back to 4000 B.C.E. have been found in Egyptian tombs.

It's All Greek to Me

The word *electricity* stems from the Greek word *elektron*, which means "amber." Why? The ancient Greeks discovered that rubbing a piece of amber (fossilized tree resin) with a cloth causes the amber (now charged with static electricity) to attract feathers.

Disappearing Chalk

What You Need

❑ 1 cup white vinegar

❑ One stick white chalk

❑ Clean, empty glass jar

What To Do

Pour the vinegar in the jar, drop in the stick of chalk, and wait ten minutes.

What Happens

Bubbles rise from the stick of chalk, which soon breaks into small pieces and dissolves completely.

Why It Works

The acetic acid in the vinegar dissolves the calcium carbonate in the chalk, releasing carbon dioxide gas.

Bizarre Facts

■ Chalk is soft, fine-grained white limestone—made mostly from small shells and calcite crystals—that did not change into hard rock.

■ Man-made classroom chalk is molded from plaster of Paris, which consists of calcium sulfate made from gypsum.

■ The White Cliffs of Dover in England are made from chalk.

■ When acid rain falls on statues containing limestone or marble (a hard crystalline metamorphic form of limestone), the statue slowly deteriorates, like the detail on the Parthenon in Athens, Greece.

■ The chalk deposits of western Kansas contain preserved skeletons of extinct sea serpents, flying reptiles, birds, and fishes.

■ Placing a piece of chalk in a jewelry box prevents rust by absorbing the moisture.

Chalk It All Up to Archeology

Most chalk was formed during the Cretaceous Period of time (beginning 130 million years ago and lasting 65 million years), named from the Latin word *creta*, meaning "chalk."

Disappearing Ink

What You Need

- Teakettle
- Water
- Measuring Cup
- Measuring spoons
- 1 teaspoon cornstarch
- Spoon
- Eyedropper
- Iodine
- Q-tips cotton swabs
- White paper

What To Do

With adult supervision, fill the teakettle with water and bring it to a boil. Carefully fill the measuring cup with one cup of boiling water. Add one teaspoon of cornstarch to the boiling water and stir with the spoon until completely dissolved. Remove any lumps. Using the eyedropper, add roughly ten drops of iodine to turn the mixture bluish and stir well with the spoon. Use a Q-tips cotton swab to write your secret message or draw a picture on a piece of white paper. Let dry. (NOTE: Iodine is poisonous if swallowed. Do not drink or taste the iodine or the ink solution.)

What Happens

After several days, the blue ink disappears from the paper.

How It Works

Adding the cornstarch to the water turns the liquid into a relatively strong base solution (pH 7.5). The iodine, when added to starch, turns blue. When the solution dries, carbon dioxide in the air mixes with the cornstarch solution, lowering the basicity enough to cause the iodine solution to vanish.

Bizarre Facts

▓ Iodine was discovered in 1811 by French chemist Bernard Courtois, who located the chemical element in seaweed.

▓ Kelp is rich with iodine.

▓ The thyroid gland in the human body produces iodine as part of a hormone called thyroxine, which controls the body's rate of physical and mental development. A shortage of iodine in the body can cause a goiter (an enlargement on the thyroid gland). To prevent iodine shortages in the body, manufacturers add iodine to salt, which they call iodized salt.

▓ Iodine was once commonly used as an antiseptic.

▓ Between 1944 and 1947, the nuclear weapons plant in Hanford, Washington, released radioactive iodine gas into the air. Cows grazed on grass contaminated by the airborne iodine, and when people drank the milk from the cows, the radioactive iodine tended to concentrate in the thyroid gland, in sufficient amounts to cause at lease some cases of cancer.

Disappearing Peanuts

What You Need

- Rubber gloves
- Safety goggles
- Protective breathing mask
- 1 quart acetone (available at a hardware store)
- Galvanized metal bucket (not plastic)
- Large box of polystyrene peanuts

What To Do

With adult supervision and wearing the rubber gloves, safety goggles, and breathing mask, carefully pour the acetone into the bucket. Pour the polystyrene peanuts from the box into the bucket, gently swirling the bucket to swish the acetone around.

What Happens

The polystyrene peanuts vanish and only a rubbery goo remains floating in the acetone.

Why It Works

The acetone acts as a solvent, dissolving the links between the molecules in the polystyrene peanuts and releasing the trapped gas in the foam, leaving behind the liquid hydrocarbon styrene used to create the peanuts.

Bizarre Facts

◼ Polystyrene is a solid plastic that results from the homopolymerization of the liquid hydrocarbon styrene.

◼ Peanuts are actually considered a legume, not a nut.

◼ Polystyrene foam—used to make disposable coffee cups, fast-food hamburger boxes, and packing peanuts—is considered environmentally hazardous because it takes up space in landfills, requires decades to decompose, and its manufacture causes the release of hazardous chemicals.

◼ Foamed polystyrene is used to make cups, bowls, plates, trays, clamshell containers, meat trays, egg cartons, and protective packaging for shipping electronics and other fragile items.

◼ Solid polystyrene is used to make cutlery, yogurt and cottage cheese containers, cups, clear salad bar containers, and video and audiocassette housings.

◼ Polystyrene foam can be recycled by turning the plastic into pea-size pellets for use in wall insulation and industrial packaging.

◼ The comic strip "Peanuts," created by Charles Schulz and featuring Charlie Brown and Snoopy, was named "Peanuts" (meaning "tiny people") by King Features Syndicate against Schulz's wishes.

Working for Peanuts

Mr. Peanut was designed in 1916 by a Suffolk, Virginia schoolchild who won five dollars in a contest sponsored by Planters Peanuts.

Dixie Cup Bridge

What You Need

❏ Seventy-five 5-ounce Dixie cups

❏ Three sheets of corrugated cardboard (16-by-16 inches)

What To Do

Place 25 paper cups upside down on the floor in five rows of five cups each. Place the first sheet of cardboard on top of the cups.

Place another 25 paper cups upside down on top of the cardboard in five rows of five cups each. Place the second sheet of cardboard on top of the cups.

Place the last 25 paper cups upside down on top of the cardboard in five rows of five cups each. Place the third sheet of cardboard on top of the cups.

Slowly step on top of the cardboard.

What Happens
The paper cups support your weight.

Why It Works
Each paper cup is a cylinder, capable of supporting up to sixteen pounds of weight. The sheet of cardboard helps distribute the weight equally across all the cups underneath.

Bizarre Facts
▪ Inventor Hugh Moore's paper cup factory was located next door to the Dixie Doll Company in the same downtown loft building. The word *Dixie* printed on the company's door reminded Moore of the story he had heard as a boy about "dixies," the ten dollar bank notes printed with the French word *dix* in big letters across the face of the bill by a New Orleans bank renowned for its strong currency in the early 1800s. The "dixies," Moore decided, had the qualities he wanted people to associate with his paper cups, and with permission from his neighbor, he used the name for his cups.

▪ In 1923, Dixie cups produced a 2½ ounce Dixie cup for ice cream, giving the ice cream industry a way to sell individual servings of ice cream and compete with bottled soft drinks and candy bars.

▪ The Dixie Cups, a popular singing trio consisting of sisters Barbara and Rosa Hawkins with their cousin Joan Johnson, sang the 1964 hit song, "Chapel of Love."

▪ While playing telephone operator Ernestine as a guest on *Saturday Night Live*, Lily Tomlin said, "Next time you complain about your phone service, why don't you try using two Dixie cups with a string?"

Edible Glass

What You Need

- Butter
- Baking sheet
- 1 cup sugar
- Heavy stainless steel or nonstick frying pan
- Large wooden spoon

What to Do

Butter the baking sheet, and place it in the refrigerator. Put the sugar in the frying pan. With adult supervision, set the pan on a burner at low heat. Stir the sugar slowly as it heats up. The sugar will slowly turn tan, stick together in clumps, and begin melting into a pale brown liquid. Continue stirring until the sugar melts into a thick brown liquid. Pour the brown liquid into the cold baking sheet. Let cool.

What Happens

The melted sugar hardens into a sheet of edible sugar glass.

Why It Works

Sugar is made of crystals, just like glass, which is made from sand.

Bizarre Facts

■ Most glass is made from a mixture of silicon dioxide (the main ingredient in sand), soda (sodium oxide), and lime (calcium oxide).

■ Silicon dioxide is one of the most inexpensive, most plentiful materials on earth.

■ Fiber optic cable is made from glass and carries far more information than the same size wire cable.

■ In Hollywood films, fake glass windows and fake glass bottles broken over the heads of movie stars were originally made from sugar. Today, they are made from a special resin.

■ Glass can be made more fragile than paper or stronger than steel.

■ Butter was probably discovered by accident. When milk is transported in containers, the agitation naturally makes the cream congeal.

Mr. Eat-It-All

Gastroenterologists confirmed that Michael Lotito (1950–2007) of Grenoble, France, had the uncanny ability to eat and digest glass and metal. During his lifetime, Lotito ate seven television sets, six chandeliers, a computer, ten bicycles, a supermarket cart, a Cessna aircraft, and a coffin. He was nicknamed Monsieur Mangetout—French for Mr. Eat-It-All. Lotito died of natural causes.

Egg in a Bottle

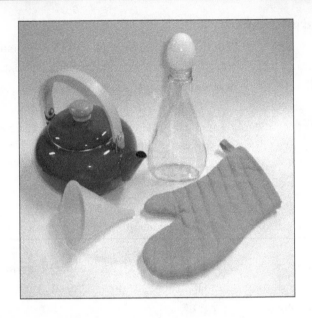

What You Need

❏ Water
❏ Teakettle
❏ Oven mitt
❏ Funnel

❏ Clean, empty, glass salad
dressing bottle
❏ Hard-boiled egg, peeled

What to Do

With adult supervision, boil water in the teakettle. Wearing the oven mitt and using the funnel, carefully fill the empty salad dressing bottle with the boiling water from the teakettle. Remove the funnel. Swirl the water around inside the bottle, and then pour the water into the sink. Quickly place the egg upright on the mouth of the bottle.

Hoax in a Bottle

A note found in a bottle on the coast of Denmark in 1946, written on a page torn from the logbook of the German U-boat *Naveclus*, and dated one year earlier, claimed that Adolf Hitler did not die in the Berlin bunker but aboard the *Naveclus*, which sank on November 15, 1945, while sailing from Finland to Spain.

What Happens

The egg is sucked into the bottle, making a very unusual sound. (To get the egg out of the bottle, hold the bottle upside down and blow into the bottle for thirty seconds. When you remove your mouth, the increased air pressure in the bottle forces the egg out of the bottle.)

Why It Works

The heat from the boiling water causes the air inside the bottle to expand, forcing some of it out. As the air begins to cool inside the bottle, it contracts, reducing the air pressure inside the bottle. The greater air pressure outside the bottle forces the egg into the bottle.

Bizarre Facts

■ The tradition of exchanging colored eggs in the springtime predates Easter by several centuries.

■ The ancient Egyptians buried eggs, a symbol of resurrection and birth, in their tombs.

■ According to *The Guinness Book of Records*, White Horse Scotch Whiskey makes the smallest bottles of liquor now sold in the world. The bottles are two inches high and contain 1.184 milliliters (less than one-eighth teaspoon).

Electric Lemon

What You Need

- Wire cutters
- Six stiff copper wires, 6-inches long, 16 grade
- Five galvanized nails
- Five lemons
- 1.8-volt red LED bulb

What to Do

With the wire cutters, strip 1 inch of insulation off both ends of each wire. Wrap one end of the first five wires to its own nail.

Squeeze the lemons, crushing them gently, to loosen the pulp inside so the juice flows inside the fruit.

Insert the nail end of the first wire into the first lemon.

Insert the bare end of the second wire into the first lemon (without letting the wire and the nail touch each other inside the lemon). Insert the nail end of that same wire into the second lemon.

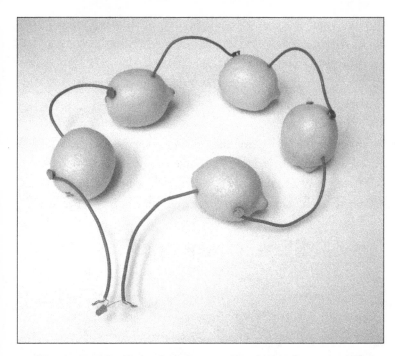

Repeat until all the lemons are wired together. Attach the remaining two wires to the LED bulb.

What Happens

The lemons light up the LED bulb.

Why It Works

The citric acid in the lemon juice acts as an electrolyte, conducting an electron flow between the copper in the wire and the bit of steel in the nail, turning each lemon into a battery. (If the LED bulb doesn't light up, add more lemons.)

Bizarre Facts

■ The modern-day household battery should be called a cell, not a battery. A *battery* is an array of single cells.

- The lemon is actually a type of berry called a *hesperidium*.
- Lemons are believed to have originated in northeastern India, near the Himalayas.
- The first lemon trees in America were planted in 1493 by Christopher Columbus.
- Actor Jack Lemmon's last name really was Lemmon.
- The word *lemon* is slang for "a defective automobile," derived from the fruit's unavoidable sour taste.
- The amount of energy needed to make a battery is fifty times greater than the amount of energy that same battery produces.

Shaking a Leg

The battery owes its discovery to frogs' legs. In the 1780s, Luigi Galvani, a professor of anatomy at Bologna University, noticed that the legs of dead frogs twitched when they were hung from hooks on a rail. Fellow professor Allesandro Volta of Pavia University deduced that the frogs' legs were completing the circuit between the copper hooks and the iron rail, prompting him to produce a Voltaic pile, the world's first battery, in 1800.

Electric Waterfall

What You Need

- Clean, empty, plastic jar with a plastic lid
- Safety goggles
- Electric drill with ¼-inch bit
- Black construction paper
- Scissors
- Black electrical tape
- Water
- Two hardcover books
- Kitchen sink
- Flashlight

What To Do

With adult supervision and wearing safety goggles, drill a hole in the top of the plastic lid, ½ inch away from the edge. Drill a second hole on the opposite side of the lid, directly across from the first hole. Cut a sheet of black construction paper so you can roll it around the side of the jar and tape it securely in place, leaving the bottom of the jar exposed. Use the black electrical tape to

55

cover up any remaining area of the jar where light might escape. Fill the jar with water and screw on the prepared lid tightly.

Turn off the lights in the room. Using the lit flashlight to watch what you are doing, lie the jar down between two books, positioning the jar so that the holes in the lid are lined up vertically and a stream of water flows out the bottom hole and into the sink. Hold the lit end of the flashlight against the bottom of the jar.

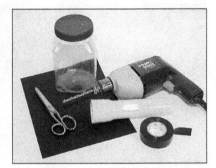

What Happens

The beam of light from the flashlight bends as it travels through the curving stream of water flowing from the hole.

Why It Works

Although rays of light travel in a straight line, rays of light can be bent (or refracted). When light enters water at an angle, it bends because the speed of light is slower in water than in air. Some of the light enters the water, and some of the light is reflected back. When the curve of the water stream is less than the critical angle of 49 degrees (as in this case), light is reflected back and forth between the surfaces of the water stream until it reaches the end and touches the sink. (The critical angle is different for different substances.) The water stream glows because some light emerges at ripples along the way.

Bizarre Facts

■ Light travels 186,282 miles per second in a vacuum.
■ Light travels from the sun to the earth in about eight minutes.

56

Not Much Left to Drink

Only about 3 percent of the earth's water is fresh, and 75 percent of that is frozen in glaciers and icecaps.

■ A pencil in a glass of water appears to be broken at the water line surface because of light refraction.

■ The amount that a ray of light bends when passing from one medium into another is called the *index of refraction*. Finding the index of refraction requires trigonometry. According to Snell's Law, developed by Dutch mathematician Willebrord Snell van Royen, the index of refraction equals the sine of the angle of incidence divided by the sine of the angle of refraction.

■ The average person consumes approximately 16,000 gallons of water in his or her lifetime.

■ The average American uses seventy gallons of water per day.

■ Water covers more than 70 percent of the earth's surface.

■ Wearing a mask underwater provides an air space so the swimmer's eyes can focus. However, when light changes speed going from water to the air inside the mask, the resulting refraction magnifies everything 25 percent, making things appear larger or closer.

■ As white light travels deeper through water, the water absorbs the colors one by one. Water absorbs red first, followed by orange, yellow, green, and blue.

Exploding Coca-Cola

What You Need

- Safety goggles
- One plastic soda bottle cap
- Electric drill with ¼-inch bit
- One pack Mentos
- Nail (1¾ inch 5d finish)
- Dental floss
- Masking tape
- One full, 2-liter bottle Coca-Cola

What You Do

With parental supervision and wearing safety goggles, drill a ¼-inch hole through the center of the plastic bottle cap. Remove the plastic insert from inside the bottle cap.

Open the pack of Mentos and take out six candies. Gently punch a hole through the center of each of the six Mentos candies by slowly pushing the nail through it, without cracking the hard candy coating. (Use the flat side of the plastic dental floss case to push the nail through the candy.)

Thread one end of the dental floss through the first Mentos candy and tie the dental floss to itself with two knots. Thread the free end of the dental floss through the remaining five Mentos candies so they hang together like beads.

Thread the free end of the dental floss through the hole drilled in the bottle cap so the Mentos candies hang from inside the cap. Pull the free end of the dental floss taut and drape it over the outer end of the bottle cap and tape in place with a 1-inch-long piece of masking tape (with one end of the tape folded back against its sticky side to create a tab).

Set the bottle of Coca-Cola on a flat surface outside and open the cap. Gently insert the strung Mentos into the neck of the bottle and screw the prepared cap securely in place without letting the Mentos touch the surface of the soda. Peel off the tape, allowing the string of Mentos candies to drop into the Coca-Cola, and quickly step back five feet.

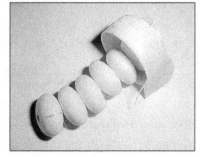

What Happens

The Coca-Cola shoots from the hole in the bottle cap, creating a geyser seven feet high for approximately ten seconds, emptying half the soda from the bottle.

Why It Works

The gelatin and gum arabic from the dissolving Mentos candies weaken the surface tension of the water in the soda, allowing the carbon dioxide bubbles to expand. At the same time, the rough surface of the Mentos candies lets new bubbles form more quickly (a process called nucleation). As more of the candies dissolve, both processes accelerate, rapidly producing foam. The resulting pressure inside the bottle forces a geyser of foam to spray from the bottle.

Bizarre Facts

■ When cooking spaghetti in a pot of boiling water, organic materials leach out from the cooking pasta and weaken the surface tension of the water in the pot. This makes it easier for bubbles and foam to form, often causing the water to boil over. Adding a drop of vegetable oil to the pot of boiling water before adding the spaghetti strengthens the surface tension of the water and prevents it from boiling over.

■ If you add a scoop of ice cream to a glass of root beer, the gums and proteins from the melting ice cream weaken the surface tension of the root beer, and the root beer foams over.

■ If you drop a Mentos candy into a glass of flat soda, nothing will happen.

■ Bookkeeper Frank M. Robinson, one of Coca-Cola inventor Dr. John Styth Pemberton's four partners, suggested naming the elixir after two main ingredients: the coca leaf and the kola nut. He suggested spelling *kola* with a *c* for the sake of alliteration.

Exploding Rubber Egg

What You Need

❑ Clean, empty glass jar with lid
❑ 2 cups white vinegar
❑ Raw egg
❑ Water
❑ Blue food coloring

What To Do

Fill the glass jar with vinegar, drop in the egg, and seal the lid. Let sit for three days or until the egg shell completely disintegrates, leaving only the membrane.

Drain all the vinegar from the jar and remove the egg, being careful not to poke the delicate membrane. If any egg shell remains, gently rub it off the membrane. Drop the egg from six inches above the tabletop. Fill the jar with water, add ten drops of blue food coloring, reseal the lid, and let sit for several days.

What Happens

The egg bounces like a rubber ball, and when left in the blue water, the egg turns blue and enlarges until the membrane bursts.

How It Works

The acetic acid in the vinegar dissolves the calcium in the egg-shell, making the shell disintegrate while leaving the membrane intact. The blue water, being a less concentrated solution than the egg's contents, passes through the semipermeable egg membrane (by osmosis), causing the pressure inside the egg to increase until the membrane bursts.

Bizarre Facts

■ English author Samuel Butler wrote: "A hen is only an egg's way of making another egg."

■ The working title of the Beatles' song "Yesterday" was "Scrambled Eggs."

■ The average ostrich egg, approximately 24 times the size of a hen's egg, can support the weight of a 280-pound human.

■ Frogs do not drink water. They absorb water into their bodies by osmosis.

■ Eggplants are named after the fact that the vegetable is shaped like a purple egg.

■ Legend holds that Simon of Cyrene, the egg merchant who helped carry Jesus's crucifix to Calvary, returned to his farm to discover that all of his hens' eggs had turned to a rainbow of colors.

■ The ancient Greeks placed eggs atop graves. When the Greeks took over ancient Israel, many Jews adopted Hellenistic practices. To this day, Jews place rocks atop gravestones to signal that the grave has been visited and the loved one remembered (perhaps as a substitute for eggs).

Eggsistentialism

On the 1960s television series *Batman*, Vincent Price played the arch-criminal Egghead, the smartest villain in the world (with an egg-shaped head), whose "eggs-cellent" vocabulary included such exclamatory phrases as "eggs-traodinary," "egg-ceptional," "eggs-hilarating," and "egg-citing."

Fake Blood

What You Need

- 1½ cups Karo light corn syrup
- ½ cup Rose's Grenadine
- Food coloring (red and blue)
- Clean, empty glass jar
- Spoon

What to Do

Mix the corn syrup, grenadine, the entire contents of a bottle of red food coloring, and three drops of blue food coloring in the jar to produce a nice, deep blood red. Stir well.

What Happens

You've made a fairly realistic prop blood for stage or screen.

Why It Works

The grenadine improves the flow of the corn syrup "blood" and lets it soak more realistically into clothing and other fabric.

Bizarre Facts

■ Hershey's Chocolate Syrup was commonly used as fake blood in black-and-white Hollywood movies.

■ Lobsters have blue blood.

■ Bloody horror movies worth missing: *Blood Orgy of the She Devils* (1972), *Blood Suckers from Outer Space* (1984), *The Blood Spattered Bride* (1974), and *Bloodsucking Pharaohs from Pittsburgh* (1991).

■ To simulate bullet shots in movies, the special effects department attaches "squibs"—small, non-metallic explosive charges—beneath the actor's clothing. Latex "blood bags," filled with a bright red, gelatin-based fluid, can be attached to the squibs, which, when detonated, burst the bags and send the fake blood flowing.

■ People who live in high altitudes have up to two quarts more blood than people who live at sea level.

■ The common vampire bat, which does attack people and drink their blood, is only three inches long.

■ In 1970, Warren C. Jyrich, a fifty-year-old hemophiliac undergoing open heart surgery at the Michael Reese Hospital in Chicago, required 1,900 pints of blood. That's roughly the blood of 190 people.

The Bloody Mary

The Bloody Mary, invented in 1920 by Ferdinand Petiot, a bartender at Harry's New York Bar in Paris, was originally named "Bucket of Blood" after a club in Chicago. Contrary to popular belief, the Bloody Mary was not named after Mary, Queen of Scots, who was beheaded by her cousin, Queen Elizabeth I. The red concoction was named after Queen Mary I of England, who was known as "Bloody Mary" for the brutal persecutions she caused the Protestants in an attempt to bring them back to the Roman Catholic faith. During her five-year reign, more than three hundred people were burned at the stake.

Falling Egg

What You Need

- ❏ Clean, empty glass jar
- ❏ Water
- ❏ Blue food coloring
- ❏ 8-inch-square piece of cardboard
- ❏ Empty aspirin bottle
- ❏ Fresh egg

What to Do

Fill the jar halfway with water and add three drops of blue food coloring. Place the piece of cardboard over the mouth of the jar. Place the aspirin bottle in the center of the cardboard directly over the center of the jar. Place the egg point first into the aspirin bottle. Hold the jar firmly with one hand, and quickly pull the piece of cardboard straight out from under the aspirin bottle.

What Happens

The aspirin bottle tumbles off and the egg drops into the jar of water.

Why It Works

As Sir Isaac Newton's first law of motion states: Objects at rest remain at rest, and objects in motion remain in motion, unless acted upon by an outside force. The stationary egg remains at rest, but once its support is pulled out from under it, gravity pulls the egg straight down into the jar of water.

Bizarre Facts

■ Newton's first law also explains how a skilled magician can pull a tablecloth out from under a fully set dinner table.

■ Eggs do not crush under the weight of a mother bird as she sits on the nest because when a force is applied to an egg, the curve of the egg distributes the force over a wide area away from the point of contact.

■ You can spin a hard-boiled egg in place on a flat surface. A raw egg will wobble.

■ In 1979, David Donoghue dropped fresh eggs from a helicopter 650 feet above a golf course in Tokyo, Japan. The eggs remained intact.

■ Dale Lyons of Meriden, Great Britain, holds the world record for running while carrying an egg on a spoon. On April 23, 1990, he ran 26 miles, 385 yards in 3 hours, 47 minutes.

■ On April 1, 2007, Cypress Gardens Adventure Park in Winter Haven, Florida, held the biggest Easter egg hunt on record in the United States using 510,000 plastic, candy-filled eggs.

■ If you hold an egg in your palm and then try to squeeze your hand into a fist, you will not crush the egg.

■ In parts of Germany during the 1880s, Easter eggs etched with the recipient's name and birth date were accepted in courts of law as birth certificates.

Laying the Golden Egg

In 1979, the College of Agriculture at the University of Missouri recorded a hen that laid 371 eggs in 364 days.

Flaming Creamer

What You Need

- Safety goggles
- Clean, empty coffee can with plastic lid
- Electric drill with ¼-inch bit
- Plastic drinking straw
- Scissors
- Black electrical tape
- Candle (1-inch in diameter)
- Matches
- 2 tablespoons Coffee-mate powdered nondairy creamer
- Fire extinguisher

What to Do

With adult supervision and wearing safety goggles, drill a hole in the side of the coffee can, as close to the bottom rim as possible.

Insert 1 inch of the straw into the hole in the side of the can. Use the scissors to cut small pieces of electrical tape to seal the spaces between the straw and the hole in the can.

Using the scissors, cut off a piece of candle 1-inch tall, mak-

67

ing sure to leave enough wick so you can light it later. Using the matches, light the bigger candle and carefully let ten drops of hot wax drip to the center of the inside of the metal bottom of the coffee can. Secure the shorter candle upright inside the coffee can on the center of the bottom.

Outdoors, place 2 tablespoons of nondairy creamer around the candle inside the coffee can. With adult supervision, wearing safety goggles, and with a fire extinguisher on hand, carefully light the candle and cover the coffee can with the lid.

Blow hard into the straw. Remember to be careful with fire!

What Happens
The plastic lid blows off the coffee can with a momentary blast of fire and smoke.

Why It Works
Powdered nondairy creamer, when suspended in air, is flammable.

Things Ain't What They Seem To Be

Some brand-name nondairy creamers contain casein, which is a dairy product.

Bizarre Facts
■ Flour is also highly flammable when suspended in air, explaining why explosions often occur in grain storage facilities.
■ Mongolian ruler Genghis Khan's armies carried rations of dried milk.
■ Early Africans milked buffaloes.
■ Powdered milk is made by evaporating all the water from pasteurized skimmed milk.

Floating Bubbles

What You Need

❑ Cardboard box
❑ Plastic garbage bag
❑ Work gloves
❑ Large block of dry
   ice (always handle
   dry ice with work
   gloves)
❑ Bottle of bubbles and
   blow ring

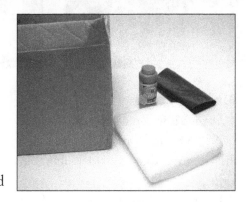

What You Do

Line the inside of the box with the garbage bag, and, wearing work gloves, place the block of dry ice in the bottom of the box. (Never touch dry ice with your bare hands since it can burn your skin if it touches it directly. Also never eat dry ice as it can be fatal.) Blow some bubbles into the box.

What Happens

The bubbles float in midair, get larger, absorb each other, and change colors.

Why It Works

The bubbles float on a cushion of carbon dioxide. The carbon dioxide is a heavy gas, the bubbles are filled with air, which is made up of much lighter gases. The bubbles change colors and turn transparent as the soap film falls to the bottom of the bubble—eventually causing the bubble to pop.

Bizarre Facts

▪ Dry ice is frozen carbon dioxide gas.

▪ When heated, dry ice does not melt into liquid form. Instead, the frozen carbon dioxide sublimes, returning directly to gas form.

▪ Dry ice is colder than ice made from frozen water. Dry ice can reach a temperature as low as −112 degrees Fahrenheit.

▪ To make dry ice, carbon dioxide gas is compressed into a liquid and then cooled and evaporated to make carbon dioxide snow, which is then compressed into blocks of solid dry ice.

▪ If ingested, dry ice can cause death due to its low temperature.

▪ Dry ice can be used to ship perishable foods because the ice does not melt into liquid (but rather sublimates into gas).

Boiling Ice

A small piece of dry ice, placed in a glass of water, will appear to make boiling water which steams. This produces an excellent "boiling caldron" effect for Halloween celebrations.

▪ Inserting a small piece of dry ice into an uninflated balloon and tying a knot in the neck of the balloon will produce enough carbon dioxide to inflate the balloon.

Floating Egg

What You Need

- Clean, empty glass jar
- Hot tap water
- Canister of salt
- Tablespoon
- Raw egg

What To Do

Fill the jar halfway with hot tap water. Stir in one tablespoon of salt at a time until no more salt will dissolve in the water. Gently drop the egg into the salt water. The egg floats in the water. Now slowly pour regular tap water over the egg, filling up the rest of the jar.

What Happens

The egg floats on top of the salt water, but remains under the layer of regular tap water.

Bad Egg

If placed in a jar of regular tap water, a fresh egg will sink to the bottom because the egg is denser than water. If the egg has gone bad, however, it will float in the water. The bad egg floats because the yolk and albumen have dried up, making the egg less dense than water.

Why It Works

Welcome to Archimedes' Principle—the explanation behind buoyancy. The ancient Greek mathematician Archimedes stated that an object placed in a fluid is buoyed upward with a force equal to the weight of the fluid it displaces. In other words, adding salt to water increases the density of the water. The salt water is denser than the egg, so the egg floats. The egg, however, is denser than regular water, so the egg sinks in it. If left undisturbed, the egg will remain floating in the middle of the jar for several days.

Bizarre Facts

■ If you let the egg sit in the jar unrefrigerated for several days, the egg will go bad and float to the surface. (A bad egg, when broken, reeks; dispose of the bad egg carefully).

■ Ancient Greek mathematician Archimedes, who discovered the laws of the lever and pulley, also invented the catapult.

■ White eggs and brown eggs are equally nutritious.

■ Hard-boiling an ostrich egg requires forty minutes.

■ In ancient Egypt, the apricot was called the "egg of the sun."

■ On September 6, 1981, in Siilinjorvi, Finland, Risto Arntikainen threw a fresh hen's egg 317 feet, 10 inches to Jyrki Korhonen, without breaking it.

■ In *Pudd'nhead Wilson*, Mark Twain wrote: "Put all your eggs in one basket and—watch that basket."

Floating Paper Clip

What You Need
❏ Scissors
❏ String
❏ Paper clip
❏ Scotch Tape
❏ Clean, empty glass jar with a
 metal lid
❏ Magnet

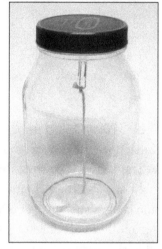

What to Do
Cut a piece of string a few inches long and tie one end to the paper clip. Tape the other end to the bottom of the jar. Place the magnet inside the lid. Show the jar to your audience with the clip lying at the bottom of the jar. Screw the lid on the jar, and then turn the jar upside down so the clip hangs from the string. Turn the jar right-side up again.

What Happens
The paper clip remains suspended in air.

Why It Works
The magnet attracts the paper clip, but the string prevents the paper clip from being pulled to the magnet.

Bizarre Facts
■ Scientists believe that the earth acts like a huge magnet with magnetic north and south poles.

A Separate Absurdity

In his nonfiction books, anthropologist Carlos Castaneda reported seeing Mexican Indian sorcerer Don Genaro fly through the air, walk horizontally up the side of a tree, and fly back. He also claimed to have seen both Don Genaro and Don Juan jump off a cliff, twirl slowly in the air, reach bottom, and float back up to the top. Many consider Castaneda's nonfiction books to be works of novelistic fancy.

■ In 1900, Cornelius J. Brosnan received a patent for the Konaclip, better known today as the paper clip.

■ The Indian rope trick (making an ordinary rope rise into the night sky and then climbing up the rope) is done by stretching thin cable some fifty feet off the ground across a valley and slinging another fine cable over it—one end held by an assistant far off in the distance, the other end attached to a small hook that is attached to the rope.

■ In the 1890s, French magician Alexander Herrmann created the illusion of levitation by a having a "hypnotized" woman lie on a board placed between the back of two chairs, taking away the two chairs, and then passing a hoop around the floating woman. The illusion is created by having an assistant behind a curtain operate a strong frame set with a "gooseneck" (a metal slat curved like the neck of a goose).

■ In an eighteenth-century treatise, Pope Benedict XIV reported that several eyewitnesses, including Pope Urban VIII, had seen St. Joseph of Cupertino rise into the air "when in a condition of ecstatic rapture."

■ In *Belgravia* magazine, Reverend C. M. Davies, a critic of psychics, detailed his own eyewitness account of seeing British psychic Daniel Dunglas Home float around a drawing room for five minutes.

Flying Ping-Pong Ball

What You Need
❑ Blow-dryer
❑ Ping-Pong ball

What You Do
With adult supervision, plug in the blow-dryer, turn it on "high cool," and aim the nozzle straight up in the air. Hold a Ping-Pong ball directly over the nozzle and gently release the ball.

What Happens
The Ping-Pong ball hovers upward in the middle of the airstream and stays in place, as if floating on a cushion of air. You can also

tilt the blow-dryer roughly 20 degrees and the Ping-Pong ball will remaining floating.

Why It Works
In 1738, Swiss mathematician Daniel Bernoulli discovered Bernoulli's principle—namely, that the pressure in a fluid drops as it moves faster. The Ping-Pong ball stays floating in the column of fast-moving air because the high pressure surrounding the column pushes the ball back into the column of low-pressure air.

Bizarre Facts

■ Bernoulli's principle explains how airplanes fly. The top of an airplane wing curves so that air flowing over it speeds up, moving faster than the air flowing below the wing.

■ Bernoulli's principle also explains how baseball pitchers throw curve balls. The pitcher spins the ball as it is thrown. The spin increases the speed of the air flowing over one side of the ball, and the disproportionate air pressure causes the ball to curve from its original course.

■ In the early 1900s, manufacturers frequently promoted multiple uses for a newly devised electrical appliance in the hopes of increasing sales of the product. An early advertisement for a vacuum cleaner called the Pneumatic Cleaner showed a woman using a hose connected to the machine to dry her hair, assuring readers that the vacuum produced a "current of pure, fresh air from the exhaust." At the time, using the exhaust from a vacuum cleaner was the only way to blow-dry hair. The handheld electric hair dryer had yet to be invented because no one had developed a motor small enough to power it—until 1922, when the advent of the blender produced the first fractional horsepower motor.

Hot Air

The development of the blow-dryer enabled hair salons to accomplish in minutes what once took hours. Thus they handled more clients in less time, increasing revenues.

■ In 1920, two companies in Racine, Wisconsin—the Racine Universal Motor Company and Hamilton Beach—essentially combined the vacuum cleaner with the blender to create the handheld hair dryer. The Racine Universal Motor Company introduced the "Race." Hamilton Beach launched the "Cyclone."

■ While the name *Ping-Pong* sounds Chinese, the game originated in England during the late 1800s.

Flying Potato

What You Need

❏ Small red rose potato
❏ Two forks
❏ Toothpick
❏ Drinking glass

What To Do

Insert the tines of one fork into one end of the potato. Insert the tines of the second fork into the potato at the other end of the potato so that the handles of the forks are at a 45 degree angle from each other.

Insert the toothpick into the potato between the two forks.

Place the end of the toothpick on the edge of the drinking glass.

What Happens

The toothpick miraculously supports the weight of the forks and potato, keeping them balanced on the edge of the drinking glass.

77

(If the potato does not balance on the toothpick, reposition the forks so the angle between the handles is smaller.)

Why It Works

The angle of the forks spreads their weight so the center of gravity of the potato, forks, and toothpick is concentrated on the toothpick.

Bizarre Facts

■ The word *fork* comes from the Latin *furca*, meaning "pitchfork."

■ Forks, usually made with only two tines, originated in Tuscany, Italy, in the eleventh century, but the clergy frowned upon their use, insisting that only human fingers should be used to eat God's bounty—along with spoons and knives.

■ Forks became popular as eating utensils among French nobility during the eighteenth century.

Small Potatoes

Shortly after World War II ended, inventor George Lerner created a set of plastic noses, ears, eyes, and mouth parts that could be pushed into fruits or vegetables to create comical food characters. Toy companies, however, rejected Lerner's new toy, convinced that American consumers, still clinging to a World War II mentality to conserve resources, would refuse to buy any toy that wasted food.

In 1952, Hasbro, Inc., a toy company based in Pawtucket, Rhode Island, launched Mr. Potato Head, a box of plastic parts (eyes, ears, noses, and mouths) that children could use to adorn a real potato (provided by Mom and Dad).

One year after introducing Mr. Potato Head, Hasbro launched Mrs. Potato Head. On February 11, 1985, after 23 years of marriage, Hasbro introduced the Potato Heads' first Tater Tot—Baby Potato Head.

Flying Rice Krispies

What You Need

- Two strips of stiff 1-by-3-inch cardboard
- Pencil
- Scissors
- Glass chimney from a hurricane lamp
- Scotch Tape
- Strip of 4-by-11-inch posterboard
- ½ cup Rice Krispies cereal
- Blow-dryer

What to Do

Holding the 1-by-3-inch cardboard strips horizontally, draw a line through the center of each strip, perpendicular to its length. Cut ½ inch into each line. Fit the two cardboard strips together at the cuts to form a sturdy, cross-shaped support for the glass chimney.

Set the glass chimney on top of the cardboard support. Tape the ends of the 4-by-11-inch strip of posterboard together to form a loop. Place the loop around the glass chimney's cardboard support to create a fence.

Pour the Rice Krispies into the opening at the top of the chimney. Use the blow-dryer to blow a stream of air across the top of the glass chimney.

What Happens

The Rice Krispies rise and float through the glass chimney.

Why It Works

The air blowing rapidly across the top of the glass chimney creates lower air pressure at the top of the glass chimney, causing the Rice Krispies to rise. Bernoulli's principle, named for Swiss mathematician Daniel Bernoulli, puts it this way: As the velocity of a gas or liquid increases, the pressure perpendicular to its direction of flow decreases.

Bizarre Facts

■ In the 1890s, the health-conscious Dr. John Harvey Kellogg and his brother, W. K. Kellogg, invented Corn Flakes while working at the Battle Creek Sanitarium in Michigan and had a spiteful relationship for the rest of their lives.

■ The Kellogg brothers invented peanut butter, but failed to patent it.

■ C. W. Post, a former patient of the Kellogg brothers at the Battle Creek Sanitarium, launched Grape Nuts, a cereal similar to granola, a cereal invented by the Kelloggs and served at the sanitarium.

■ Although he invented granola and Corn Flakes, Dr. John Harvey Kellogg breakfasted daily on seven graham crackers.

■ Bernoulli's principle makes it is possible to sail a boat forward against the wind.

Snap! Crackle! Pop!

In 1928, the Kellogg Company introduced Kellogg's Rice Krispies as "the Talking Cereal." In 1933, a year after the phrase "Snap! Crackle! Pop!" was printed on the box, a tiny, nameless gnome wearing a baker's hat appeared on the side of the box. He eventually became known as Snap, and in the mid-1930s he was joined by Crackle (wearing a red striped stocking cap) and Pop (in a military hat). Snap, Crackle, and Pop are the Kellogg Company's oldest cartoon characters.

Foam Factory

What You Need

- Measuring cup
- 1 can shaving cream
- Large spoon
- Large glass bowl
- 4-ounce bottle of Elmer's Glue-All

What To Do

Fill the measuring cup with one cup shaving cream. Spoon the shaving cream into the mixing bowl. Empty the bottle of Elmer's Glue-All into the mixing bowl. With the spoon, whip the glue and shaving cream together. Mold whatever you want out of the mixture. Let set overnight to dry.

What Happens

The mixture solidifies into foam.

Why It Works

Whipping the shaving cream and the glue together essentially fills the polyvinyl acetate molecules in the glue with air-filled soap lather. The glue dries filled with millions of tiny cells of air, much like foam rubber.

Bizarre Facts

■ In 1920, Frank B. Shields, a former MIT chemistry instructor, developed the formula for Barbasol, one of the first brushless shaving creams on the market. Shields developed Barbasol especially for men with tough beards and tender skin because he had both of those shaving problems. The white cream in a tube—providing a quick, smooth shave—immediately won the allegiance of thousands, eliminating the drudgery of having to lather up shaving soap in a mug with a shaving brush and then rubbing it onto the face.

■ The original Barbasol factory and offices were both located in a small second-floor room in downtown Indianapolis. The tubes were filled with shaving cream, clipped, and packaged by hand. At the most, only thirty or forty gross made up an entire day's production schedule.

■ A 1937 advertisement for Barbasol read, "Barbasol does to your face what it takes to make the ladies want to touch it."

■ Shaving in the shower wastes an average of 10 to 35 gallons of water. To conserve water, fill the sink basin with an inch of water and vigorously rinse your razor often in the water after every second or third stroke.

■ According to archaeologists, men shaved their faces as far back as the Stone Age—20,000 years ago. Prehistoric men shaved with clam shells, shark teeth, sharpened pieces of flint, and knives.

■ The longest beard, according to *Guinness Book of Records*, measured 17.5 feet long and was presented to the Smithsonian Institution in 1967.

Close Shave

In 1931, Charles Goetz, a senior chemistry major at the University of Illinois, worked part-time in the Diary Bacteriology Department, improving milk sterilization techniques. Convinced that storing milk under high gas pressure might inhibit bacterial growth, Goetz began experimenting—only to discover that milk released from a pressurized vessel foamed. Realizing that cream would become whipped cream, Goetz began seeking a gas that would not saturate the cream with its own bad flavor. At the suggestion of a local dentist, Goetz succeeded in infusing cream with tasteless, odorless, nonflammable nitrous oxide, giving birth to aerosol whipped cream and aerosol shaving cream.

■ Ancient Egyptians shaved their faces and heads before hand-to-hand combat so the enemy had less to grab. Archaeologists have discovered gold and copper razors in Egyptian tombs dating back to the fourth century B.C.E.

■ Aerosol cans to deliver shaving cream were introduced in the mid-1950s.

■ The first shaving creams specifically targeted to women were introduced in 1986.

■ Seventy percent of women rate clean-shaven men as sexy.

■ The product known today as Elmer's Glue-All was first introduced by Borden in 1947 under the brand name Cascorez, packaged in two-ounce glass jars with wooden applicators. Sales did not take off until 1951 when Borden chose Elsie the Cow's husband, Elmer, as the marketing symbol. In 1952, Borden repackaged Glue-All into the familiar plastic squeeze bottle with the orange applicator top.

Fried Marbles

What You Need

- Safety goggles
- Bag of transparent marbles
- Frying pan
- Oven mitt
- Wooden spoon
- Large pot
- Cold water
- Ice

What to Do

With adult supervision and wearing safety goggles, place the marbles in the frying pan, and, wearing the oven mitt, heat them over a high heat, stirring with the wooden spoon.

Fill the large pot with cold water, add the ice, and let cool.

When the marbles are piping hot, carefully pour them into the ice water. Let them cool off, and then dry.

What Happens

The glass inside the marbles shatters into shards and looks like shimmering crystal.

Why It Works

When glass goes from extreme heat to extreme cold, it cracks from the inside out.

Bizarre Facts

■ Although marbles have been made from clay, stone, wood, glass, and steel, most marbles today are made from glass.

■ Marbles found in ancient Egypt and Rome can be seen in the British Museum.

■ In 1846, German glassblower Elias Greiner invented a tool called marble scissors, making the manufacture of glass marbles economically feasible.

Losing Your Marbles?

The most common method of shooting a marble is known as *fulking*.

■ World War I cut off the supply of marbles to North America.

■ Most glass marbles in the United States are made at a plant in Clarksburg, West Virginia, which makes millions every year.

■ A variety of colors and intricate patterns create a wide range of glass marbles, including the Immy, Moonstone, Rainbow, Marine, Cat's Eye, Genuine Carnelian, First American, Japanese Cat's Eye, Scrap Glass, and Peppermint Stripe.

■ Marble games include Archboard, Bounce About, Bounce Eye, Conqueror, Die Shot, Dobblers, Eggs in the Bush, Handers, Hundreds, Increase Pound, Lag Out, Long Taw, Odds or Even, One Step, Picking Plums, Pyramid, Ring Taw, Spanners, and Three Holes.

■ Marbles are often used at the bottom of clear glass vases to support flowers or at the bottom of fish tanks.

Green Pennies

What You Need
- Three coffee filters
- Bowl
- White vinegar
- Ten pennies

What To Do
Place the three coffee filters, stacked together, in the bowl. Wet the coffee filters with the vinegar. Place the pennies on the coffee filter. Let sit for 24 hours.

What Happens
The pennies turn green.

Why It Works
The acetic acid in the vinegar mixes with the copper coating on the pennies to form copper acetate, which appears green in color.

Bizarre Facts

■ The Statue of Liberty appears green because the acid in rain turns the copper statue into cooper acetate.

■ A penny weighs more than the Reddish Hermit Hummingbird and the Bee Hummingbird.

■ The bumblebee bat of Thailand, the world's smallest mammal, weighs less than a penny.

■ "A bad penny" is someone undesirable, "a pretty penny," is a large sum of money, and "an honest penny" is money earned honestly.

> **A Penny for Your Thoughts**
>
> Pennies are copper-coated zinc alloyed with 2.5 percent copper.

■ Seventh-century King Penda of the Anglo-Saxon kingdom of Mercia developed the silver penny and named the coin after himself.

■ The female name Penny is a variation on the more formal name Penelope.

■ A "penny pincher" is a cheapskate, a "penny wheep" is a small glass of beer," and a "penny ante" is a paltry sum of money.

Green Slime

What You Need

- ❏ 4-ounce bottle of Elmer's Glue-All
- ❏ Two large glass bowls
- ❏ Water
- ❏ Green food coloring
- ❏ Large spoon
- ❏ Measuring cup
- ❏ 1 teaspoon 20 Mule Team Borax
- ❏ Ziploc Storage Bag or airtight container

What to Do

Empty the bottle of Elmer's Glue-All into the first bowl. Fill the empty glue bottle with water and then pour it into the bowl of glue. Add ten drops of food coloring and stir well.

88

In the second bowl, mix the borax with 1 cup water. Stir until the powder dissolves.

Slowly pour the colored glue into the bowl containing the borax solution, stirring as you do so. Remove the thick glob that forms, and knead the glob with your hands until it feels smooth and dry. Discard the excess water remaining in the bowl. Store the Green Slime in the Ziploc Storage Bag or airtight container.

What Happens

The resulting soft, pliable, rubbery glob snaps if pulled quickly, stretches if pulled slowly, and slowly oozes to the floor if placed over the edge of a table.

Why It Works

The polyvinyl acetate molecules in the glue act like invisible chain links drifting around the water. The borax molecules (sodium tetraborate) act like little padlocks, locking the chain links together wherever they touch the chain. The locks and chains form a interconnected "fishnet," and the water molecules act like fish trapped in the net.

Bizarre Facts

■ Green Slime is a non-Newtonian fluid—a liquid that does not abide by any of Sir Isaac Newton's laws on how liquids behave. Quicksand, gelatin, and ketchup are all non-Newtonian fluids.

■ Increasing the amount of borax in the second bowl makes the slime thicker. Decreasing the amount of borax makes the slime more slimy and oozy.

■ A non-Newtonian fluid's ability to flow can be changed by applying a force. Pushing or pulling on the slime makes it temporarily thicker and less oozy.

■ 20 Mule Team Borax is named for the twenty-mule teams used during the late nineteenth century to transport borax 165 miles across the desert from Death Valley to the nearest train depot in Mojave, California. The twenty-day round trip started 190 feet below sea level and climbed to an elevation of over 4,000 feet before it ended.

■ Between 1883 and 1889, the twenty-mule teams hauled more than 20 million pounds of borax out of Death Valley. During this time, not a single animal was lost nor did a single wagon break down.

■ Today it would take more than 250 mule teams to transport the borax ore processed in just one day at Borax's modern facility in the Mojave desert.

■ Although the mule teams were replaced by railroad cars in 1889, twenty-mule teams continued to make promotional and ceremonial appearances at events ranging from the 1904 St. Louis World's Fair to President Woodrow Wilson's inauguration in 1917. They won first place in the 1917 Pasadena Rose Parade and attended the dedication of the San Francisco Bay Bridge in 1937.

■ According to legend, borax was used by Egyptians in mummification.

■ In the furniture business, the word *borax* signifies cheap, mass-produced furniture.

■ 20 Mule Team Borax was once proclaimed to be a "miracle mineral" and was used to aid digestion, keep milk sweet, improve the complexion, remove dandruff, and even cure epilepsy.

Homemade Chalk

What You Need

- Empty cardboard toilet paper tubes
- Scissors
- Waxed paper
- Scotch Tape
- Plaster of Paris
- Warm water
- Bowl
- Spoon
- Tempera paints in various colors

What To Do

For each stick of chalk you wish to make, line the inside of an empty toilet paper tube with waxed paper and seal one end with tape.

With a spoon, mix two parts plaster of Paris with one part

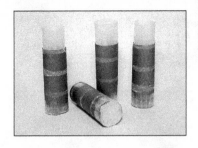

warm water in a bowl and add approximately two spoonfuls of tempera paint to achieve the desired color. Pour the mixture into the prepared toilet paper tubes. Gently tap the tube to release air bubbles from the plaster. Repeat for each color of chalk you wish to make.

Let the plaster mixture dry for 48 hours. Peel off the cardboard tube and wax paper.

What Happens

You've made colored chalk.

How It Works

Chalk is made from limestone, but man-made classroom chalk is molded from plaster of Paris, which consists of calcium sulfate made from gypsum.

Bizarre Facts

■ Calcium carbonate, the main ingredient in chalk, is the key ingredient in toothpaste and antacid tablets.

■ Plaster of Paris is made by heating crushed gypsum (hydrated

Don't Have a Cow, Man

On the animated television comedy series *The Simpsons*, Bart Simpson was forced to write on a blackboard one hundred times at school, "I will not waste chalk."

calcium sulfate) until it has been dehydrated, leaving behind a white powder. Adding water to the powder causes a chemical reaction, changing the powder back to gypsum, which sets hard when the water evaporates.

■ The word *gypsum* is Latin for "chalk."

■ Plaster of Paris is naturally fire retardant. At roughly 600 degrees Fahrenheit the water molecules stored in the plaster ooze out. This is why walls appear "sweaty" after a fire.

■ Ants will not cross a thick chalk line.

Homemade UFO

What You Need

❑ Scissors
❑ Six plastic drinking straws
❑ Scotch Tape
❑ Ruler
❑ One tall, clear kitchen
     trash bag, 16 gallon, 24
     inches by 33 inches by
     0.31 mil (available as
     Simply Value brand)
❑ Spool of thread
❑ 4-by-4-inch square of alu-
     minum foil
❑ Indelible marker
❑ Four birthday candles
❑ Barbecue butane lighter
❑ Fire extinguisher
❑ Safety goggles

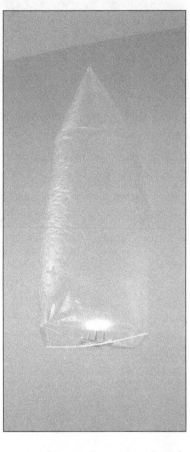

What to Do

Using the scissors, snip off the
shorter end of the bendable
straw that includes the accordion-type hinge. Snip two ¼-inch
slits in one end of a straw, insert the slit end into one end of a second straw, and secure in place with a small strip of Scotch Tape.
Repeat until you make a length of straw 17 inches long (or 70
percent of the width of the plastic trash bag). Repeat this process
to create a second, matching length of straw.

94

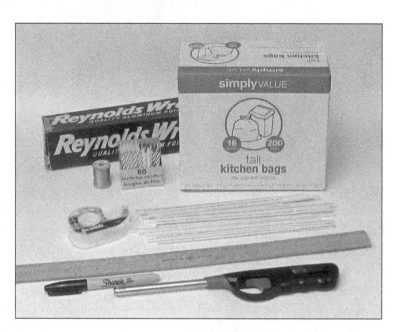

Cross the two lengths of straw to create an X, and secure the two straws in place with a piece of Scotch Tape.

Tie the end of the thread from the spool to the spot where the straws cross. The thread will serve as a leash for the UFO.

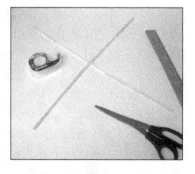

Using the ruler, measure 1 inch in from the midpoint of each side of the square of aluminum foil and make a dot with the indelible marker for a total of four dots. With adult supervision, use the barbecue butane lighter to melt two drops of hot

95

wax from the end of one birthday candle onto one of the dots on the aluminum foil, and stand the candle upright in the hot wax, holding it in place until the wax dries. Repeat this process to secure each one of the remaining three candles on the remaining three dots.

Place the aluminum candle platform upright in the center of the straw crossbeam so that the corners of the aluminum foil square touch the straws and tape in place to the straws from underneath the aluminum foil. (Make sure the spots where the candles are mounted to the aluminum foil do not sit on top of the straws; otherwise, the heat from the candle flames will melt the straws).

Open the plastic trash bag and carefully tape the four ends of the straws to the inside edges of the rim of the trash bag, so that the candles face inside the trash bag.

With adult supervision, a fire extinguisher on hand, and wearing safety goggles, have a partner hold the bottom of the trash bag three feet off the ground and use the barbecue butane lighter to carefully ignite the four candles, without letting the flames touch or get near the sides of the plastic bag. Hold out the sides of the plastic bag so the hot air produced by the candles does not melt the plastic together. Allow the plastic bag to fill with hot air and then release the bag.

Should a lit candle accidentally fall off the aluminum foil, the drop though the air will, in most cases, extinguish the flame. The sides of the plastic bag, if touched by flame, will melt without catching fire; however, be cautious of droplets of hot plastic.

What Happens

As hot air rises inside the plastic bag, the bag fully inflates and rises. The plastic bag drifts upward until the flames go out and air inside the plastic bag cools. If launched at night, the flame also illuminates the balloon, making it look like a UFO.

Why It Works

Heat rises. The hot air inside the plastic bag, lighter than the surrounding cool air, causes the plastic bag to rise.

Bizarre Facts

▧ Weather balloons are commonly mistaken for UFOs.

▧ The Bible includes the prophet Ezekiel's eyewitness account of an aerial chariot containing four-winged monsters and traveling on wheels within wheels in a stormy wind with flashes of fire. UFO enthusiasts believe Ezekiel was describing an alien space craft. Biblical scholars point out that Ezekiel describes each monster as having four faces unique to this planet: the face of a man, a lion, an ox, and an eagle.

▧ When viewed from the sky, the Nazca lines—the gigantic markings made centuries ago on the plains of Nazca, Peru—can be clearly seen as enormous drawings of animals. While UFO enthusiasts suggest that these fantastic drawings were meant as signals for extraterrestrial visitors, archaeologists have shown that these drawings were ancient astronomers' method of tracking the constellations for agricultural purposes.

■ Seven percent of all Americans claim to have seen a UFO.

■ In the liner notes to his album *Walls and Bridges*, John Lennon claimed: "On the 23rd Aug. 1974 at 9 o'clock, I saw a U.F.O."

■ The "live long and prosper" hand gesture made by Mr. Spock on *Star Trek* is the hand gesture used by Jewish priests (*kohanim*) while saying certain prayers.

■ David Prowse, the actor dressed as Darth Vader in the 1977 movie *Star Wars*, spoke all of Vader's lines as the movie was filmed. He did not know until he saw a screening of the movie that his voice had been dubbed over by James Earl Jones.

No Place Like Home

In the 1939 movie *The Wizard of Oz*, the Wizard claims "I'm an old Kansas man myself," yet he takes off from the Emerald City in a hot-air balloon painted with the words "Omaha State Fair," which could only take place in Nebraska—proving that he too is filled with hot air.

Hose Phone

What You Need
❑ Hacksaw
❑ 100-foot-long rubber garden hose
❑ Two large funnels
❑ Black electrical tape

What to Do
Use the hacksaw to cut off the metal couplers at the ends of the garden hose. Insert the narrow end of a funnel into each end of the hose and secure in place with black electrical tape. Hold one funnel to your ear and the second funnel to your mouth. Say something. You will hear your voice with a slight delay. Stretch the hose across the yard and have a friend speak into one funnel while you listen with the second funnel.

What Happens
The hose works just like a phone, with a slight delay.

Why It Works
At sea level, sound waves travel approximately 760 miles per hour. That means sound travels through a 100-foot-long hose in roughly $\frac{1}{10}$ of a second.

Bizarre Facts
■ Bats make high-frequency sounds while flying and navigate using the echoes from these sounds to determine the distance and

direction of objects in the area. Using reflected sound waves to navigate is called *echolocation*.

■ In 1860, the inventor of vulcanized rubber, Charles Goodyear, died, having failed to perfect a practical use for his invention and leaving his family with nearly $200,000 in debts. Ten years later, Dr. Benjamin Franklin Goodrich, determined to cash in on rubber's untapped potential, founded the B. F. Goodrich Company in Akron, Ohio, and began producing the world's first rubber hoses.

■ The 1970s television sitcom *Welcome Back, Kotter*, starring Gabe Kaplan and John Travolta, popularized the meaningless catchphrase "Up your nose with a rubber hose."

■ Around 350 B.C.E., the Greek philosopher Aristotle suggested that sound is carried to our ears by the movement of air. He was wrong. Sound travels in waves.

■ The higher the altitude, the slower sound travels.

■ Two days before he broke the sound barrier, Chuck Yeager broke two ribs.

■ On Alexander Graham Bell's telephone, the mouthpiece also served as the earpiece. Thomas Edison separated the transmitter from the receiver, making the telephone easier to use.

Telegram about the Telephone

In 1876, an internal memo at Western Union read: "This 'telephone' has too many shortcomings to be seriously considered as a means of communication. The device is inherently of no value to us."

Human Light Bulb

What You Need

❑ Wool sweater

❑ Fluorescent light bulb

What to Do

Put on the wool sweater. In a dark room, rub the fluorescent light bulb briskly against the sweater.

What Happens

The fluorescent light bulb glows.

Why It Works

The friction creates a static charge strong enough to cause the gas inside the tube to fluoresce.

Striking

American park ranger Roy Sullivan was struck by lightning seven times between 1942 and 1977.

101

Bizarre Facts

■ On the 1960s television show *The Addams Family*, electrically-charged Uncle Fester makes a light bulb illuminate by simply placing it in his mouth.

■ The fluorescent bulb is more economical and energy efficient than the incandescent bulb, which wastes up to 80 percent of its energy generating heat.

■ While Thomas Edison is credited with inventing the first incandescent lamp using carbon for the filament, English inventor Joseph Swan patented his incandescent lamp using carbon for the filament in 1878, a year before Edison.

■ When cartoon characters get an idea, an incandescent light bulb goes off over their heads—rarely a fluorescent bulb.

■ Thomas Edison, nicknamed "the Wizard of Menlo Park," was expelled from school in Port Huron, Michigan, after the schoolmaster incorrectly diagnosed him as being mentally retarded. Edison was actually partially deaf, the result of a bout with scarlet fever.

■ Lightning travels between one hundred and one thousand miles per second, generating a temperature up to 54,000 degrees Fahrenheit, almost five times hotter than the surface of the sun.

Hydrogen Balloon

What You Need

❏ Funnel
❏ 2 cups water
❏ Clean, empty wine bottle
❏ Rubber gloves
❏ Safety goggles
❏ 3 tablespoons Crystal Drano

(sodium hydroxide)
❏ Aluminum foil (12-by-12 inches)
❏ Balloon
❏ String

What to Do

Working outdoors with adult supervision, use the funnel to pour the water into the wine bottle. Wearing rubber gloves and safety goggles, carefully add the Crystal Drano into the bottle. Carefully swirl the bottle to dissolve the Drano. Make twenty small balls from the aluminum foil (about ½ inch in diameter) and drop them into the bottle. Stretch the balloon well and immediately place the neck of the balloon over the mouth of the bottle. Let it inflate as big as it can get. This takes about ten minutes.

If the solution doesn't give off enough gas to fill the balloon, add more aluminum; if the glass gets too hot, you've used too much aluminum.

When the balloon is full, tie it off with the string. (Do not breathe the vapors from the bottle, and carefully dispose of the remaining solution from the bottle.)

What Happens

A chemical reactions gives off hydrogen gas, which fills the balloon. Since hydrogen is lighter than air, the balloon floats.

Why It Works

The aluminum interacts with the lye in the Crystal Drano, causing the hydrogen molecules to separate from the sodium hydroxide and water.

Bizarre Facts

■ A box of Reynolds Wrap—the only nationally distributed brand of aluminum foil—can be found in three out of four American households.

■ The *Hindenburg*, one of the largest airships ever built, burst into flames in 1937 over Lakehurst, New Jersey, when its hydrogen filled bag exploded, killing 36 people. Today airships are filled with helium.

■ Hydrogen is used as rocket fuel because the combustion reaction between hydrogen and oxygen propels the exhaust gas (primarily water vapor) out of the rocket's engine at 7,910 miles per hour—creating the enormous thrust that lifted the 4.5 million-pound space shuttle into orbit.

■ The first hydrogen bomb, detonated in Enewetak Atoll in the Pacific in 1952, used fission to cause the fusion of the nuclei of two hydrogen atoms—yielding an explosion equivalent to ten million tons of TNT.

■ If a hydrogen atom were the size of a golf ball, a golf ball would be the size of the earth.

Unsolved Mystery

On June 30, 1908, an explosion as powerful as a hydrogen bomb allegedly shook the Tunguska region of Siberia. Some witnesses reported seeing a fireball or a mushroom cloud. In 1927, Soviet scientists examined the 2000-square-kilometer site and determined that the charred earth had not been caused by a meteorite. Although many scientists today believe the crater was caused by an asteroid or comet, no one knows for certain.

Ice Cream Machine

What You Need

- Mixing bowl
- Box of instant pudding (to make six ½-cup servings)
- 3 cups milk
- Whisk
- Small, clean, empty coffee can with lid (net weight 11.5 ounces)
- Black electrical tape
- Large, clean, empty coffee can with lid (net weight 34.5 ounces)
- Ice
- Rock salt

What To Do

Empty the contents of the pudding mix into the mixing bowl. Add 3 cups milk (according to the directions on the back of the box), and mix well using the whisk.

Pour the pudding solution into the small, clean coffee can. Secure the plastic lid in place and use the electrical tape to make the lid watertight.

Place the small coffee can into the large coffee can. Fill the rest of the large can with ice up to the top of the small can. Fill the rest of the space with rock salt. Secure the plastic lid in

place and use the electrical tape to make the lid watertight.

Take the can outside and roll it back across the lawn or patio for fifteen minutes. Bring the can back inside, peel the tape from the lid of the large can, pour out the melted ice and salt, and refill with fresh ice and fresh salt. Secure the lid in place again, and roll the can outside for another fifteen minutes.

Bring the can back inside, peel off the tape from the larger can, pour out the melted ice and salt, and wash off the smaller can with tap water in the sink. Dry the can. Store in the freezer for twelve hours. Peel off the tape from the smaller can, remove the lid, and scoop the contents into bowls.

What Happens

You've made ice cream.

Why It Works

Mixing ice with salt in the compartment around the small can creates freezing temperatures, causing the mixture inside the small can to freeze. Rolling the large can causes the smaller can to roll around in the ice. As the smaller can rolls, air bubbles are whipped into the ice cream, increasing the volume of the mix.

Bizarre Facts

■ No one knows who first invented ice cream or when. In the late 1500s, Europeans used ice, snow, and saltpeter to freeze mixtures of cream, fruit, and spices.

■ Almost all ice cream was made at home until 1851, when Baltimore milk dealer Jacob Fussell established the first ice cream factory.

■ The edible ice cream cone, invented by Italo Marchiony of Hoboken, New Jersey, was first served at the 1904 World's Fair in St. Louis, Missouri.

■ The most popular flavor of ice cream in the United States is vanilla, accounting for approximately one-third of all the ice cream sold in the country. The second most popular flavor is chocolate, followed by strawberry.

■ The United States produces more than 1.5 billion gallons of ice cream, ice milk, sherbet, and water ice every year.

■ Approximately 10 percent of all the milk produced in the United States is used to make ice cream and other frozen desserts.

■ As mayor of Carmel, California, actor Clint Eastwood's first act was to legalize ice cream parlors.

■ Americans eat more ice cream than do the people of any other nation in the world.

■ The average American eats roughly 14.5 quarts of ice cream in a year.

We All Scream for Ice Cream

On July 24, 1988, Palm Dairies Ltd. of Alberta, Canada, created the worlds' largest ice cream sundae—made from 44,689 pounds, 8 ounces of ice cream; 9,688 pounds, 2 ounces of syrup; and 537 pounds, 3 ounces of topping.

Ice Cube Saw

What You Need

- Plastic Rubbermaid shoe box container
- Water
- Wire cutters
- 36-inch length of 22-gauge cooper wire
- Two clean, empty bleach bottles with caps
- Tall, plastic garbage pail
- Plastic children's wading pool

What To Do

Fill the plastic shoe box container with water and place in a freezer for 24 hours to make a giant ice cube.

Using wire cutters, strip the coating of plastic insulation off the wire, leaving only copper wire. Tie one end of the wire to the handle of the first bleach bottle. Tie the other end of the wire to the handle of the second bleach bottle. Fill the two bleach bottles with water and seal tightly.

In a cool, shady spot, place the garbage pail upside down in the center of the plastic wading pool. Carefully remove the giant ice cube from the plastic shoe box container and place it on top of the upside down garbage pail.

Pick up the two bleach bottles (now wired together), guide the wire over the ice, and gently lower the bottles so they hang on the two sides of the garbage pail.

Let sit for several hours.

What Happens

The wire cuts through the ice cube, but the ice cube refreezes above it.

Why It Works

The pressure and the heat created by the friction of the wire from the weight of the bottles melts the ice. A small amount of water escapes the pressure by rising above the wire, but the cold temperature from the ice cube refreezes the water.

Bizarre Facts

- The average iceberg weighs 20 million tons.
- The white trail emitted from a plane is actually ice.
- Rap singer Ice Cube's real name is O'Shea Jackson, and rap singer Vanilla Ice's real name is Robert Van Winkle.
- In 1851, surgeon John Gorrie of Apalachicola, Florida, built the first commercial ice-making machine.
- As water freezes into ice, the water molecules move apart and form a rigid pattern of crystals, expanding in volume by about one-eleventh.
- Ice floats because expansion makes it lighter than water.

The Iceman Cometh

Approximately one-tenth of the earth's surface is permanently covered with ice. If all the ice melted, the sea would rise by about two hundred feet and many of the world's largest cities—including New York, Los Angeles, London, and Tokyo—would be underwater.

Inflating Glove

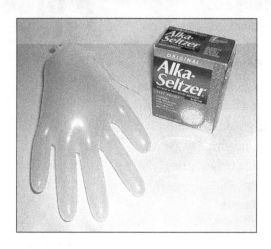

What You Need

❏ Dental floss
❏ Latex glove

❏ Water
❏ Four Alka-Seltzer tablets

What To Do

Tie a small loop in the end of a 12-inch length of dental floss. Thread the free end of the dental floss through the loop, creating a slipknot large enough to encircle the open end of the glove. Set aside.

Carefully fill the four fingers of the glove with water, leaving the thumb empty. Carefully slip four Alka-Seltzer tablets inside the dry thumb. Place the prepared slipknot over the open end of the glove, and then pull the end of the dental floss to tighten the knot and securely close the end of the glove. Wrap the dental floss tightly around the slip knot several times, and then tie another knot to secure in place.

111

Hold the thumb of the glove and guide each one of the four Alka-Seltzer tablets into its own water-filled finger.

What Happens
The glove inflates like a balloon.

Why It Works
When activated in water, the Alka-Seltzer tablets release carbon dioxide gas, filling the sealed glove. With nowhere to go, the carbon dioxide inflates the glove, which expands like a balloon.

A Glove Story
During the Middle Ages, a knight would attach a woman's glove to his helmet to demonstrate love or devotion.

Bizarre Facts
■ German Kaiser Wilhelm II frequently attempted to conceal his withered arm by posing with his hand resting on a sword or by holding gloves.

■ When terrorists sent letters filled with anthrax spores through the United States Postal Service in the wake of the attacks on the World Trade Center and Pentagon on September 11, 2001, sales of latex gloves skyrocketed.

■ During the Middle Ages, throwing down a glove signalled a challenge to a duel and was called "throwing down the gauntlet." Whoever picked up the glove, accepted the challenge.

■ Soldiers from every country in the world salute with their right hand.

■ The Japanese word *karate* means "empty hand."

■ The three major candidates in the 1992 Presidential election—George Bush, Bill Clinton, and Ross Perot—were all left-handed.

■ William Shakespeare's father was a glovemaker.

Invisible Inks

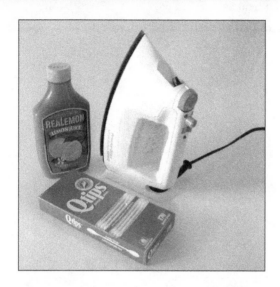

Lemon Juice or Milk

What You Need
- Q-tips cotton swab
- Lemon juice or milk
- Paper
- Electric iron
- Ironing board

What to Do
Using the Q-tips cotton swab as a paintbrush and the lemon juice or milk as ink, write your message on the paper. Let dry at room temperature. With adult supervision, iron the piece of paper.

What Happens
The invisible message appears.

Why It Works

Lemon juice and milk are simple organic liquids that appear invisible once they have dried on a sheet of paper, but darken when held over a heat source.

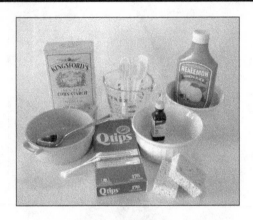

Cornstarch

What You Need

- ❑ 1 teaspoon cornstarch
- ❑ Measuring cup
- ❑ Water
- ❑ Microwave-safe container
- ❑ Spoon
- ❑ Q-tips cotton swab
- ❑ Paper
- ❑ Two small bowls
- ❑ Iodine
- ❑ Two small sponges
- ❑ Eyedropper
- ❑ Lemon juice

What to Do

Mix the cornstarch and ½ cup water in the microwave-safe container. Stir until smooth. Heat in the microwave on high for fifteen seconds, stir, and then heat on high for 45 seconds more.

Dip the Q-tips cotton swab into the mixture. Write your message on the paper. Let dry.

In one small bowl, add ten drops of iodine to ¼ cup water. Sponge the iodine solution lightly over the paper.

What Happens

The message appears in dark blue. (If the paper contains starch the paper may also turn light blue.) In the second bowl, pour ¼ cup lemon juice and sponge it over the message. The message will disappear again.

Why It Works

The iodine reacts with the cornstarch to form a new compound that appears blue-black. The ascorbic acid in lemon juice combines with the iodine to form a colorless compound.

Bizarre Facts

▨ Prisoners of war used their own sweat and saliva as invisible ink.

▨ India ink was actually invented in China.

▨ During World War II, a German spy named Oswald Job was executed in 1944 after a crystal of concentrated invisible ink was found hidden in a hollow key that he carried in his pocket.

▨ A message in invisible ink, written beneath a one-sentence love letter on a postcard sent from Poland in 1943, vividly describes the horrific conditions in a Nazi death camp and makes an urgent request for supplies. The postcard is on display at the Yad Vashem Holocaust Museum in Jerusalem.

Close Shave

In 500 B.C.E., the ancient Greek tyrant Histiaeus wrote a secret message on the shaved head of a slave, waited for his hair to grow back, and then sent the slave across enemy lines. When the slave's head was shaved, the message was revealed.

Kaleidoscope

What You Need

- Ruler
- Pencil
- Cardboard mailing tube 2 inches in diameter with plastic end caps (available at an office supply store)
- Serrated knife
- Three pieces of mirror, each 1½-by-10 inches (available at a glass company)
- Black electrical tape
- Paper towels
- Acetate sheet (from the cover of a clear folder)
- Indelible marker
- Scissors
- Rubber cement
- 14-ounce frosted plastic drinking cup (available at a supermarket)
- Colored plastic drinking straws
- Safety goggles
- Electric drill with ½-inch bit
- Sandpaper
- Wrapping paper
- Clear laminating plastic

What To Do

Measure a 10½-inch length of the mailing tube, draw a circle around the tube as a guideline, and with adult supervision, use a serrated knife to cut the length from the mailing tube.

Using electrical tape, carefully cover all four edges of each of the three pieces of mirrored glass to prevent cuts from the sharp edges. Avoid taping the mirrored surfaces.

Carefully place the mirrors together to form a triangle, with the mirrored surfaces on the inside of the triangle. Making certain the edges are even, tape the mirrors together along the edges and then run strips of tape around the entire triangle.

Slide the triangle of mirrors into the mailing tube aligning one end of the triangle with one end of the tube—leaving ½ inch of free space at the other end of the tube.

Crumple up small pieces of paper towel and, using the pencil, shove them between the walls of the mailing tube and outsides of the mirrors to cushion the mirrors and wedge them snugly inside the tube.

Lay a plastic end cap from the mailing tube on the sheet of acetate, and use the indelible marker to trace a circle around the end cap. Using a pair of scissors, cut one disk from the acetate.

On the end of the mailing tube that is even with the mirrors, paint rubber cement along the cardboard edge of the mailing

117

tube. Let dry for a few minutes, and then apply a second coat of rubber cement. Fit the acetate disk over the end of the tube, press tightly, and then stand the tube on end so the disk is pressed against the floor or tabletop, and let dry.

Using scissors cut the bottom from the 14-ounce frosted plastic cup roughly ¼ inch above the base (so that the piece can fit over one end of the mailing tube like a cap and be secured in place). Cut ¼-inch pieces from the colored drinking straws. Place the straw pieces into the plastic cup bottom. Fit the piece over the acetate disk on the end of the mailing tube. Hold the tube up to the light, and look through the other end while turning the tube. If you do not see changing designs, add more pieces of colored straws.

When you are content with the number of straws in the cup, remove the cup, apply rubber cement to make a ¼-inch band around the perimeter of the mailing tube, and replace the plastic cup, holding firmly until the glue sets.

With adult supervision and wearing safety goggles, drill a ½-inch hole in the center of the plastic end cap of the mailing tube. Use fine sandpaper to smooth the edges of the hole.

Cut a circle 1 inch in diameter from the sheet of acetate, and glue the disk over the hole on the inside of the plastic end cap. Place the prepared end cap over the open end of the mailing tube to create a peephole.

Cut a piece of wrapping paper to fit around the tube, and adhere to the tube with rubber cement. Then cover the wrapping paper with clear laminating plastic.

Look through the peephole, hold the other end of the kaleidoscope up to the light, and gently turn the tube.

What Happens
You see changing symmetrical patterns of bright color.

Why It Works
The colors at the end of the tube are reflected in the mirrors, creating multiple images. Turning the tube shifts the colored shapes, creating new patterns.

Bizarre Facts
■ Sir David Brewster invented the kaleidoscope in 1815 and patented his invention in 1817.

■ Designers have used kaleidoscopes to find new patterns for fabrics, rugs, and wallpapers.

■ In the children's story "Snow White and the Seven Dwarves," as told by the Brothers Grimm, the wicked queen asks her magic mirror, "Mirror, mirror, on the wall, who's the fairest of them all?"

■ In the song "Lucy in the Sky with Diamonds," the Beatles sing about a girl named Lucy with kaleidoscope eyes.

Kaleidoscopic Quote
In 1818, Dr. Peter M. Roget (who created his thesaurus in 1852) praised the kaleidoscope in *Blackwood's Magazine*: "In the memory of man, no invention, and no work, whether addressed to the imagination or to the understanding, ever produced such an effect."

Killer Straw

What You Need
❑ Raw potato
❑ Two plastic drinking
   straws

What to Do
Place the potato on a ta-
ble. Hold the first straw
at the top (without cov-
ering the hole) and try
to stab it into the potato.

Hold your thumb over the hole in the top of the second straw and
try to stab it into the potato.

What Happens
The open-ended straw bends, and only a bit of the straw pen-
etrates the potato. The closed straw cuts deeply into the potato.

Why It Works
The air trapped inside the straw gives the straw enough strength
to penetrate the skin of the potato. As the straw enters the potato, the potato plug compresses the air inside the straw, increasing
the air pressure, and strengthening the straw.

Bizarre Facts
■ During colonial times, New Englanders, convinced that raw
potatoes contained an aphrodisiac which induced behavior that
shortened a person's life, fed potatoes to pigs as fodder.

■ After serving as ambassador to France, Thomas Jefferson brought the recipe for French-fried potatoes to America, where he served them to guests at Monticello, popularizing French fries in the United States.

■ The potato originated in South America, where the Incas cultivated and crossbred it. In the 1500s, while Spanish explorers introduced the potato to Europe, English explorers brought the potato to the United Kingdom, where it became the principal crop of Ireland. Today, China grows approximately 23 percent of the world's potatoes, more than any other country.

■ Dan Quayle, vice-president of the United States under George Bush, corrected a student in a spelling bee, insisting that the word *potato* is spelled *potatoe*. It isn't.

■ Ira Gershwin wrote the famous lyric, "You like potato, and I like po-tah-to" in the 1937 hit song "Let's Call the Whole Thing Off," written in collaboration with his brother George Gershwin.

■ The phrases "that's the last straw" and "the straw that broke the camel's back," originated with Archbishop John Bramhall, who, in 1655, wrote: "It is the last feather that breaks the horse's back."

Grasping at Straws

Straw and hay are not the same thing. Straw is the dried stems of grains such as wheat, rye, oats, and barley. Hay is dried grasses or other plants.

Lava Lamps

Color Globs Lava Lamp

What You Need
- Clean, empty 1-liter bottle
- Funnel
- ¾ cup of water
- 1 liter of vegetable oil
- Food coloring (red, blue, or green)
- Alka-Seltzer tablets
- Jumbo flashlight

What to Do
Using the funnel, pour the water into the empty bottle. Still using the funnel, fill the rest of the bottle with vegetable oil, leaving one inch of air below the neck of the bottle. Let the liquids sit until the oil and water separate completely into two layers.

Add ten drops of either red, blue, or green food coloring into the neck of the bottle. Observe as the drops of food coloring sink through the layer of oil, come to rest on the surface of the water, and then, after a few minutes, burst through to the water.

122

Crack three Alka-Seltzer tablets in half and drop them into the bottle. Turn on the jumbo flashlight, stand it upright, and carefully place the bottle on the plastic faceplate covering the light bulb. Observe.

When the action ceases, you can add more broken tablets of Alka-Seltzer, or cap the bottle tightly and save the lava lamp for future use.

What Happens

A continuous stream of colored bubbles rise to the top of the bottle and slowly sink back down to the bottom again.

Why It Works

The water and oil are immiscible, meaning they do not mix together. The water falls to the bottom because it is more dense than the oil. The food coloring, being water soluble, sinks to the bottom, breaks though the surface tension of the water, and mixes with the water.

The Alka-Seltzer reacts with the water, creating bubbles of carbon diox-

ide gas, which, being less dense than both the water and oil, rise to the surface, pushing globules of colored water up with them. At the surface, the carbon dioxide escapes, and globules of colored water sink back down through the oil to the layer of water.

Poor Man's Lava Lamp

What You Need
❑ Bottle of club soda
❑ Four or five raisins

What to Do
Open the bottle of club soda, drop in the raisins, and reseal the cap tightly on the bottle. (The effect will also work in a drinking glass filled with club soda.)

What Happens
The raisins sink to the bottom of the bottle, slowly rise to the surface, and sink back down again, repeatedly.

Why It Works
Carbon dioxide bubbles from the club soda accumulate in the wrinkles of the raisins, eventually lifting the raisins to the surface. There the bubbles escape, and the raisins sink to the bottom again to repeat the cycle. The effect lasts longer in a sealed bottle of club soda because less carbon dioxide is able to escape.

Bizarre Facts

■ Craven Walker (1918–2000), a native of Singapore, came up with the idea for the lava lamp while drinking in a pub in Dorset, England, after World War II, and spent the next fifteen years developing it. Walker launched the "Astro lamp" in 1963, just in time for the psychedelic sixties.

■ Over 400,000 lava lamps are made each year.

■ Firewalkers in Hawaii walk barefoot over hot lava.

■ The eruption of Laki volcano in 1783 in southeast Iceland created a lava flow some 43½ miles long, the longest in recorded history.

■ Erupting lava, the molten rock that pours out of volcanoes, can reach temperatures up to 2,192 degrees Fahrenheit—more than ten times hotter than the boiling point of water.

■ The eruption of the Icelandic volcano Eyjafjallajökull on April 14, 2010, spewed forth a vast cloud of ash as high as 36,000 feet, compelling aviation authorities to ground more than 107,000 flights over the North Atlantic. Thirteen months later, on May 21, 2011, the Grímsvötn volcano under the Vatnajokull glacier in southeast Iceland spewed forth a vast cloud of ash more than 65,500 feet high, prompting aviation authorities to cancel nine hundred flights.

Magic Candle

What You Need

- Candle
- Matches
- Bowl
- Four pennies

- Clean, empty glass jar
- Blue food coloring
- 1 cup of water
- Spoon

What To Do

With adult supervision, melt the bottom of the candle and secure it upright in the center of the bowl.

Place four pennies equidistant from each other around the candle so the jar can sit over the candle with the jar rim resting on the pennies. Add three drops of blue food coloring to the water and mix well with the spoon.

Pour the colored water into the bowl, light the candle, and set the jar over the candle so it sits on the pennies.

What Happens

The water rises roughly one-fifth of the way up the jar, the flame goes out, and the water remains in the jar.

Why It Works

Fire consumes the oxygen in the air. As the candle flame consumes the oxygen in the jar, the resulting pressure sucks the water into the jar to replace the lost oxygen. When the lit candle consumes all the oxygen in the jar, the flame goes out.

Bizarre Facts

■ Air is composed of 21 percent oxygen, 78 percent nitrogen, and 1 percent trace elements.

■ To human beings, air is invisible, odorless, and tasteless.

■ People have lived more than a month without food and more than a week without water, but people cannot live more than a few minutes without air.

■ Air circulators in the Holland and Lincoln Tunnels under the Hudson River, connecting New Jersey and New York, circulate fresh air through the tunnels every ninety seconds.

■ Before the invention of strings of electric Christmas lights, people decorated their Christmas trees with lit candles.

■ Flamboyant pianist Liberace was known for his elaborate grand piano decorated with rhinestones and an ornate candelabra.

Burning the Candle at Both Ends

In Sweden, on the morning of St. Lucia Day (celebrated on December 13) the oldest daughter in the home dresses in white, wears a wreath with seven lit candles on her head, and serves her family coffee and buns in bed.

Magic Crystal Garden

What You Need

- Safety goggles
- Five charcoal briquets
- Hammer
- 2-quart glass bowl
- Clean, empty glass jar
- 1 tablespoon ammonia
- 6 tablespoons salt
- 6 tablespoons Mrs. Stewart's Liquid Bluing (available from a grocery store, or visit www.mrsstewart.com)
- 2 tablespoons water
- Food coloring

What to Do

Wearing safety goggles, break up the charcoal briquets into 1-inch chunks with the hammer, and place the pieces in the bowl.

In the jar, mix the ammonia, salt, bluing, and water thoroughly. Pour the mixture over the charcoal in the bowl.

128

Sprinkle a few drops of food coloring over each piece of charcoal.

Let the bowl sit undisturbed in a safe place for 72 hours.

What Happens

Fluffy, fragile crystals form on top of the charcoal, and some climb up the sides of the bowl. To keep the crystals growing, add another batch of ammonia, salt, bluing, and water.

Why It Works

As the ammonia speeds up the evaporation of the water, the blue ion particles in the bluing and the salt get carried up into the porous charcoal, where the salt crystallizes around the blue particles as nuclei. These crystals are porous, like a sponge, and the liquid below continues to move into the openings and evaporate, leaving layers of crystals.

Bizarre Facts

■ All solids have an orderly pattern of atoms, which is repeated again and again. This orderly pattern, called *crystallinity*, can be seen in simple crystals because their shapes reveal their particular atomic structure to the naked eye.

■ Some New Age enthusiasts believe that wearing a crystal—usually an amethyst, rose quartz, or clear quartz—around the neck attracts good vibrations and can be used to better arrange a person's spiritual and physical energies. There is no scientific evidence to support this pseudoscience.

■ Crystals grow by attracting the atoms of similar compounds, which join together in an orderly pattern. Impure atoms can invade the atomic structure of the crystal and create mixed crystals of dazzling hues.

■ Some scientists theorize that birds have a tiny magnetic crystal in their brain, enabling them to navigate during migration by detecting the earth's magnetic field.

■ Crystal gardens became popular during the Depression and are still known to some as a "Depression flower" or "coal garden."

■ In 1921, Henry Ford, eager to find a use for the growing piles of wood scraps from the production of his Model Ts, learned of a process for turning the wood scraps into charcoal briquets, and one of his relatives, E. G. Kingsford, helped select the site for Ford's charcoal plant. In 1951, Ford Charcoal was renamed Kingsford Charcoal.

■ Mrs. Stewart's Bluing, a very fine blue iron powder suspended in water, optically whitens white fabric. It does not remove stains or clean the fabric, but merely adds a microscopic blue particle to white fabric that makes it appear whiter. The brightest whites have a slight blue hue that, unfortunately, washes out over time. Adding a little diluted bluing to the rinse cycle gives white fabrics this blue hue again, making them look snow-white.

■ Freshly cut carnations placed in a vase with a high content of Mrs. Stewart's Bluing in the water will by osmosis carry the blue color into the tips of the petals quickly.

Charcoal on the Moon

On July 20, 1969, Neil Armstrong, the first man on the moon, spoke the first words on the moon: "That's one small step for man, one giant leap for mankind." The second thought he expressed was: "The surface is fine and powdery. It adheres in fine layers, like powdered charcoal, to the soles and sides of my foot."

Magnetic Balloon

What You Need
❏ Handful of Rice Krispies   ❏ Wool sweater
❏ Balloon

What to Do
Pour the Rice Krispies on a tabletop. Inflate the balloon and knot it. Rub the balloon against the wool sweater, about five strokes. Hold the balloon an inch above the Rice Krispies.

What Happens
The Rice Krispies hop up and stick to the balloon.

Why It Works
The balloon rubs electrons off the wool sweater, giving the balloon a negative charge. The negative charges on the balloon attract the positive charges of the Rice Krispies, overcoming the force of gravity.

Bizarre Facts

■ The Kellogg Company is the world's largest maker of ready-to-eat cereals, selling nearly one out of every three boxes of cereal in the United States.

■ The Kellogg Company makes twelve of the top fifteen cereals in the world, including All-Bran, Froot Loops, Kellogg's Corn Flakes, Rice Krispies, and Special K.

■ In 1986, the Kellogg Company stopped giving tours of its Battle Creek, Michigan, factory to prevent industrial spies from unearthing its secret recipes.

■ In front of the Kellogg factory in Battle Creek stands a giant statue of Tony the Tiger.

■ When W. K. Kellogg refused to buy his grandson's process for making puffed corn grits (developed on company time), the younger Kellogg—John L. Kellogg—started his own company to make the new cereal. His grandfather sued, and John Kellogg committed suicide in his Chicago factory.

Kellogg's Lucky Number

W. K. Kellogg was obsessed with the number seven. Born the seventh son to a seventh son on the seventh day of the week on the seventh day of the month with a last name seven letters long, Kellogg always booked hotel rooms on the seventh floor and insisted that his Michigan license plates end in a seven—which fails to explain the introduction of Kellogg's Product 19. It is not known whether Kellogg drank 7-Up.

Magnetic Ping-Pong Ball

What You Need

❑ Ping-Pong Ball
❑ 1-foot length of dental floss

❑ Scotch Tape
❑ Kitchen sink

What To Do

Attach one end of the dental floss to the Ping-Pong ball with a piece of Scotch Tape. Turn on the water and, holding the free end of the string, let the ball hang as close as possible to the stream of water running from the faucet.

What Happens

The ball moves into the stream of running water, even if you hold the string on an angle.

Why It Works

According to Bernoulli's principle (named for Swiss mathematician Daniel Bernoulli), the water streaming rapidly creates a low pressure zone, and the full atmospheric pressure sends the ball toward it.

Bizarre Facts

■ Jim Henson made the original Kermit the Frog from a sleeve of his mother's winter coat, using two Ping-Pong balls for eyes.

■ Cubans desperate to flee the communist country have used Ping-Pong paddles as oars to raft to Miami.

■ Table tennis was originally played with paddles made from cigar box lids and balls made from champagne corks.

■ America's table-tennis team trained for the 1992 Olympics in Barcelona with a $50,000 robot that could simulate the styles of the best Ping-Pong players in the world and shoot a ball at up to sixty miles per hour.

■ No American has ever won the men's world singles championship title in table tennis.

Bouncing Back

In the late 1880s, English engineer James Gibb, determined to come up with a game he could play indoors to get some exercise during the winter and rainy weekends, invented table tennis—a miniature version of tennis. Originally marketed under the name Gossima, table tennis did not take off until 1901, when a British manufacturer of table tennis equipment renamed the game Ping-Pong.

Marble Run

What You Need

- Ruler
- Pencil
- 10-inch length of pine-wood (1-by-1 inch)
- Safety goggles
- Electric drill with ¼-inch bit
- Sandpaper
- Four identical drinking glasses
- Karo light corn syrup
- Corn oil
- Red wine vinegar
- Shampoo
- Four identical marbles

What To Do

Using the ruler and pencil, mark off every 2 inches along the 10-inch length of wood. With adult supervision and wearing safety goggles, drill a ¼-inch hole through the center of each mark in the wood. Use sandpaper to smoothen the wood.

135

Line up the four drinking glasses in a row. Fill each glass with a different liquid to the same level, roughly 1 inch from the lip of the glass. Rest the 10-inch length of wood across the tops of the glasses so that one long edge of the wood sits over the center of each glass, with a hole positioned near the center of each glass.

Rest a marble over each of the four holes on the wood. Tilt the wood toward the center of the drinking glasses so the four marbles roll into the glasses at the same time and from the same height. Observe the marbles.

What Happens

The four marbles sink to the bottom of the liquids at different speeds. The liquid containing the marble that hits the bottom first is the least viscous.

Why It Works

Different liquids possess different viscosities. The longer a marble takes to sink in a liquid, the higher that liquid's viscosity.

Bizarre Facts

■ For an interesting variation on this experiment, gently warm all four liquids equally to see how temperature affects viscosity.

■ As you pour corn syrup, it forms coils due to its high viscosity.

■ Viscosity is a measure of a liquid's thickness or resistance to flow. The thicker a liquid, the higher the viscosity. Viscosity is a type of internal friction.

■ Pushing an object into a viscous liquid increases the viscosity, making the liquid resist more.

Mutual Attraction

In liquids, the molecules, while attracted to each other, are attracted less strongly than the molecules in solids and can move around. Freezing temperatures turn liquids into solids because the lower temperature causes the molecules to move slower and attract each other. Hot temperatures turn liquids into gases because the molecules move faster and have less attraction to each other.

Marbled Paper

What You Need

- Newspaper
- Toothpicks
- Two pieces of corrugated cardboard, 3-by-8 inches (with the grooves along the 8-inch side)
- Shaving cream
- Baking sheet
- Tempera paints (different colors)
- Water
- Six paper cups
- Sheets of white paper

What to Do

Cover the workspace with newspaper. Insert toothpicks every ½ inch into the grooves along the 8-inch side of one of the pieces of corrugated cardboard to create a large comb.

Spread a layer of shaving cream about 1-inch thick onto the

baking sheet. Make the shaving cream level by spreading it out using the second piece of cardboard as a scraper. Thin different colors of tempera paint with water in paper cups, and then pour different patterns on top of the layer of shaving cream.

Use the cardboard comb to swirl the paint on top of the shaving cream (without pushing it deep into the shaving cream).

Gently lay a piece of white paper on top of the design, press down lightly, and lift off. Place the paper with the shaving cream side up on a table, and use a piece of cardboard to squeegee off the excess shaving cream. Set to dry on a sheet of newspaper for 24 hours. You can reuse the layer of shaving cream several times, and then refresh when necessary.

What Happens

Swirling patterns of color adhere to the paper.

Why It Works

The shaving cream provides a liquid surface-tension medium upon which the suspended paint floats, whirls, and spirals. The paper absorbs the marble-like pattern onto its surface.

Bizarre Facts

■ Marbling paper was practiced in Japan and China as early as the twelfth century. According to a Japanese legend, the gods gave knowledge of the marbling process to a man named Jiyemon Hiroba as a reward for his devout worship at the Katsuga Shrine in Nara in 1151.

■ For centuries, paper marbling masters worked in secrecy to maintain a shroud of mystery to prevent others from mastering the craft and going into business for themselves.

Mayonnaise Madness

What You Need
❑ Jar of mayonnaise
❑ Freezer

What To Do
Peel off the paper label from the mayonnaise jar. Place the jar of mayonnaise in the freezer overnight. In the morning, remove the jar from the freezer. Place the jar on a counter at room temperature and let it defrost. Shake the jar for five minutes.

What Happens
In the freezer, the mayonnaise turns into large chunks of white curds floating in yellow oil. No matter how hard you shake the jar, the mayonnaise remains chunks of white curds in yellow oil.

Why It Works
When frozen, mayonnaise separates and curdles because the emulsion breaks down. As mayonnaise freezes, the oil globules expand and break away from the film that surrounds them. The oil congeals and rises to the surface, and no amount of mixing can make the mayonnaise look appetizing again.

Bizarre Facts
■ Mayonnaise is a combination of the word *Mahón*, a town on the island of Minorca, and *-aise*, the French suffix for *-ese*.

■ In the eighteenth century, the Duc de Richelieu discovered a Spanish condiment made of raw egg yolk and olive oil in the port town of Mahón on the island of Minorca, one of the Balearic Islands. He brought the recipe for "Sauce of Mahón" back to France, where French chefs used it as a condiment for meats, renaming it *mayhonnaise*.

■ In 1912, Richard Hellmann, a German immigrant who owned a delicatessen in Manhattan, began selling his premixed mayonnaise in one-pound wooden "boats," graduating to glass jars the following year. Hellmann's eventually extended its distribution from the East Coast to the Rocky Mountains. Meanwhile, Best Foods, Inc., had introduced mayonnaise in California, calling it Best Foods Real Mayonnaise and expanding distribution throughout the West. Eventually the two companies merged under the Best Foods, Inc. banner, but since both brands of mayonnaise had developed strong followings, neither name was changed.

■ Hellmann's Real Mayonnaise and Best Foods Real Mayonnaise are essentially the same, although some people find Hellmann's mayonnaise slightly more tangy.

■ Mayonnaise provides essential fatty acids which the body cannot manufacture and vitamin E. It also aids in the absorption of vitamins A, D, E, and K.

■ An unopened jar of commercial mayonnaise is microbiologically safe to eat for an indefinite period of time—as is an opened jar that is refrigerated.

Poor Man's Penicillin?

The acid ingredients in mayonnaise—vinegar and lemon juice, along with salt—destroy many bacteria. Studies at the University of Wisconsin have shown that commercial mayonnaise retards the growth of salmonella in salads inoculated with the bacterium.

141

Milk Paint

What You Need

- 1½ cups Carnation NonFat Dry Milk powder
- ½ cup water
- Food coloring or pigment (available at an art supply store)
- Blender
- Paintbrush

What To Do

Mix the Carnation Nonfat Dry Milk powder and water until it is the consistency of paint. Blend in your choice of food coloring or pigment until you attain the desired hue. Thin the paint by adding more water, thicken the paint by adding more powdered milk. Brush on as you would any other paint. Let the first coat dry for at least 24 hours before adding a second coat. Let the second coat dry for three days.

What Happens
The milk is paint.

Why It Works
The milk is an emulsion that suspends the pigment.

Paint the Town Red

Barns are traditionally painted red because early farmers mixed milk paint with blood from slaughtered animals to achieve the red color.

Bizarre Facts

■ Early American colonists made their milk paint from the milk used to boil berries, resulting in an attractive gray color.

■ Milk paint is extremely durable.

■ To strip milk paint, apply ammonia, allow it to dry for about four days, and then apply bleach. Make sure you are stripping the paint in a well-ventilated area.

■ A cow gives nearly 200,000 glasses of milk in her lifetime.

■ As early as the Stone Age, people mixed pigments in animal fat to make waterproof paints.

■ Pigments are ground from minerals, earths, plants, and animals. For instance, ultramarine is ground from the semiprecious stone lapis lazuli, vermilion is ground from mercuric sulfide from volcanic rocks, ocher is ground from iron oxide, and brazilwood is made from shavings of Brazilwood.

■ The ancient Egyptians developed paints made from egg whites mixed with pigment. Artists during the Italian Renaissance used paints made from ground pigments suspended in linseed oil.

■ In tomb paintings, the ancient Etruscans depicted women in white and men in red.

■ During his lifetime, Vincent Van Gogh sold only one painting: *Red Vineyard at Arles*.

Monster Bubbles

What You Need

- Two 90-fluid-ounce bottles of Ultra Dawn dishwashing liquid
- Six 4-ounce bottles of glycerine (available at a drugstore)
- 1 gallon water
- Plastic bucket
- 12-inch-square piece of cardboard
- Plastic wading pool
- Plastic holder from a six-pack of soda cans
- Two plastic drinking straws
- String
- Wire clothes hanger
- Hula hoop
- An 18-inch-square plastic container
- Swimming mask

What To Do

Add the dishwashing liquid, water, and glycerine in the bucket. Swirl the ingredients gently to mix them without creating soapsuds. Cover the bucket with the cardboard square and let the mixture sit undisturbed for five days.

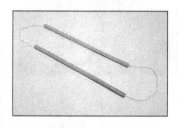

Thread a 3-foot length of string through two plastic drinking straws as if beading a necklace. Knot the ends of the string together and glide the knot inside one of the two straws.

Bend the wire clothes hanger into a circle with a handle.

Tie four 12-inch pieces of string to make four handles equidistantly around a hula hoop.

Place the plastic wading pool on a flat surface in the shade and away from any wind. Fill the pool with enough bubble solution to reach a depth of one inch.

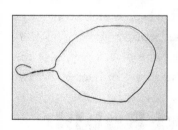

Dip the plastic holder from a six pack of soda cans into the solution and use it to blow bubbles.

Hold the two plastic drinking straws apart so the string is taut, dip it in the bubble solution, and then lift up while simultaneously bringing the straws together.

Submerge the wire clothes hanger into the bubble solution, and then lift up, swishing through the air.

145

Place the hula hoop in the wading pool. Place the 18-inch-square plastic container in the center of the pool. Have an assistant put on the swimming mask (to avoid getting soap in his or her eyes) and stand inside the plastic container in the pool. With a second assistant, slowly lift the hula hoop from the pool and over the first assistant's head.

What Happens

You create dozens of extra-strength bubbles with the plastic holder from the six pack of soda cans; you make large bubbles with the string of straws; you make enormous, monster bubbles with the wire coat hanger; and you create a bubble around a person with the hula hoop.

Why It Works

A soap bubble is a drop of water that has been stretched out into a sphere by using soap to loosen the magnetic attraction that exists between water molecules. Glycerine helps give the walls of the bubble strength. When you wave the wire coat hanger through

Bursting Your Bubble

Making monster bubbles successfully depends upon several variables, including air temperature and humidity. The more humidity in the air, the easier it is to make large bubbles. Bubbles also tend to burst quickly in direct sunlight.

the air, for instance, the air pushes apart the molecules in the soapy film, but the molecules, attracted to each other, contract, forming the smallest surface possible to contain the largest volume of air possible—a sphere.

Bizarre Facts

■ Bubbles filled with carbon dioxide (exhaled from your mouth) last longer than bubbles filled with air.

■ Bubbles made from a warm soapy solution last nearly twice as a long as bubbles made from a cold soap solution. Warmth sustains the surface tension of the bubbles. You can warm the bubble solution by placing it in a pot and heating it to 120 degrees Fahrenheit.

■ The more detergent used to make the bubble solution, the larger the bubbles will be. If you use more detergent than water (as instructed above), you can create monstrous bubbles.

■ Bubbles blown on a rainy day last longer than bubbles blown on a sunny day due to the moisture in the air.

■ By wetting one end of a plastic straw in bubble mix, you can gently push it through a large bubble and then blow a second bubble inside the first bubble.

■ The hula hoop gets it name from the Hawaiian hula dance because the gyrations made while rotating the hoop around one's waist match the movements made by doing the hula dance, originally a religious dance performed to promote fertility.

What's All the Hoopla?

The hula hoop craze swept across the United States in 1958, and within six months Americans purchased 20 million Hula hoops. The hula hoop had actually been invented in ancient Egypt three thousand years earlier, where it was made from grapevines that had been dried and stripped.

Musical Wineglasses

What You Need

- 3-foot length of pinewood (¾-by-6 inches)
- Green felt (1-by-4 feet)
- Staple gun and staples
- Scissors
- Two strips of ¼-inch wood (1 inch-by-3 feet)
- Two strips of ½-inch-thick foam (3 feet-by-1 inch) (available at a fabric store)
- Elmer's Glue-All
- Safety goggles
- Electric drill with ¹⁄₁₆-inch bit
- Eight identical wineglasses
- Screwdriver
- Six ¾-inch drywall screws
- Measuring cup
- Water
- Food coloring
- Vinegar
- Small bowl

What To Do

Wrap the felt around the wooden board and use the staple gun to secure the felt to the back of the board, wrapping the corners neatly. Using the scissors, trim any excess felt so the board rests level on a tabletop.

Glue a foam strip along the length of one side of each one of the wooden strips. Let dry.

With adult supervision and wearing safety goggles, drill a ¹⁄₁₆-

inch hole in the center and 1 inch from both ends of both foam-backed wooden strips.

Line the glasses in a row on the felt-covered board, spacing each glass 1 inch apart. Place the foam-backed wooden strips lengthwise on the felt-covered board, one strip on each side of the row of glasses—just slightly overlapping the base of each glass, with the foam-backed side facing down.

Use drywall screws to attach the wooden strips to the felt covered board to hold the glasses in place.

Fill the first wine glass with 1 ounce of water, the second glass with 2 of ounces water, the third glass with 3 ounces of water, and continue this progression until you fill the last glass with 8 ounces of water. (You may need to add more or less water to each glass to create a tuned one-octave scale.) Use the food coloring to color each glass of water a different color.

Pour 2 ounces of vinegar into the bowl, dip the index finger of each hand into the vinegar, and then rub your finger around the rim of the wine glasses in a circle to create sounds.

What Happens

By rubbing your fingers along the rims of the glasses you create a strange, ringing tone in each glass.

Why It Works

The friction from your finger causes the glass to vibrate and the resulting longitudinal sound waves resonate vertically in the glass itself and also travel around the circumference of the glass. The more water in the glass, the lower the frequency (or note) of the sound waves traveling through the glass. The ability to create different pitches enables us to tune different glasses to different musical notes on the scale, creating a musical instrument.

Bizarre Facts

■ For an interesting variation on this experiment, fill long-necked bottles with increasing amounts of water, and then blow over the bottles to make musical sounds, creating a pan flute.

■ Different pitches of an instrument are actually different frequencies of sound. High-pitched sounds are high-frequency wavelengths, meaning the sound waves travel closer together and are more numerous. Low-pitched sounds are low-frequency wavelengths, meaning the sound waves travel further apart and are less numerous.

■ In 1887, American inventor Thomas Alva Edison made the first sound recording. He recorded his voice reciting the words "Mary had a little lamb."

■ The eardrum is a thin membrane of skin approximately ⅖ inch in diameter stretched taut across the auditory canal like the skin of a drum. The eardrum is so thin that even the tiniest sound wave vibrates it to and fro. As the eardrum vibrates, it swings a tiny bone called the hammer across another bone called the anvil. The anvil shakes a third bone called the stirrup, amplifying the vibrations to the cochlea, coiled tubes resembling a snail shell, filled with liquid, and lined with minute hairs. Sound waves traveling through the fluid bend the hairs which in turn stimulate tiny nerve fibers that send the signals along the cochlear nerve to the brain, which interprets the impulses as sounds.

■ The stirrup bone is the smallest bone in the body, tinier than a grain of rice.

■ The eardrums of a frog are on the outside of its head. Behind each eye is a large exposed disk.

■ A cricket's eardrums can be seen on the side of each front leg.

Mystery Bottle

What You Need

❑ Clean, empty 1-liter
   plastic soda bottle
   with cap
❑ Water
❑ Blue food coloring
❑ Glass mixing bowl

What To Do

Fill the bottle with water,
add five drops blue food
coloring, screw on the cap
securely, and shake well to
mix up the coloring. Re-
move the cap.

Place the bottle upside down inside the bowl. A little water
will come out from the bottle. Raise the bottle 1 inch. Observe.
Raise the bottle another inch. Observe again. Raise the bottle yet
another inch. Observe.

What Happens

The level of the water in the bowl will rise only to the height of
the mouth of the bottle. The rest of the water remains inside the
bottle and does not pour out into the bowl.

Why It Works

The force pushing down on the surface of the water in the bowl
(atmospheric pressure) is greater than the force pushing the water

downward in the neck of the bottle, preventing the water from escaping from the bottle. When the mouth of the bottle is raised above the surface of the water in the bowl, air gets into the bottle, allowing water to escape. This is the theory behind pet feeders that allow water to run from a reservoir as the pet drinks water from a bowl.

Bizarre Facts

■ Temperature affects air pressure. Hot air expands, creating low air pressure. Cold air contracts, causing high air pressure.

■ The weight of air from the top of the atmosphere to the layers below creates air pressure. At sea level, the air pressure is 14.7 pounds per square inch. At the top of Mount McKinley (an altitude of 20,320 feet), the air pressure is 6.8 pounds per square inch.

■ Drinking through a straw utilizes air pressure. Sucking on the straw creates reduced air pressure inside the straw, and the greater air pressure pushing down on the surface of the liquid in the drinking glass or bottle pushes the liquid up through the straw.

■ Ancient peoples made the first bottles from animal skins. Archaeologists believe that people discovered how to blow bottles from molten glass on the end of a hollow iron pipe in the first century B.C.E.

Paper Cup Boiler

What You Need

- Fire extinguisher
- Candle
- Match
- Saucer
- Dixie cup
- Water
- Tongs

What To Do

With adult supervision and a fire extinguisher on hand, melt the bottom of the candle and secure it upright in the center of the saucer. Place the saucer on a cement surface outdoors. Light the candle.

Fill the cup halfway with water. Using tongs, hold the water-filled cup over the open flame, keeping the flame focused under the direct center of the cup (without making contact with the outer bottom rim).

What Happens
The water in the cup boils without the cup catching fire.

Why It Works
The temperature at which water boils (212 degrees Fahrenheit) is significantly lower than the temperature at which paper burns (451 degrees Fahrenheit), so the water cools the paper enough to prevent it from burning. (If the flame comes in contact with the bottom rim of the Dixie cup that does not make contact with the water, the rim and the cup may catch fire).

Bizarre Facts
■ The novel *Fahrenheit 451* by Ray Bradbury, titled for the temperature at which paper burns, is about a futuristic fireman who sets fire to banned books. The novel was made into a 1967 movie directed by François Truffaut and starred Oskar Werner and Julie Christie.

■ Davidson, Saskatchewan, is home to the world's largest coffeepot, measuring 24 feet tall and capable of holding 150,000 eight-ounce cups of coffee.

■ The phrase "my cup runneth over" first appears in the Hebrew Bible in Psalm 23.

Wet Your Whistle

The ceramic cups once used in British pubs had a whistle baked into the handle so patrons desiring a refill could whistle for service, giving birth to the phrase "Wet your whistle."

Paper Helicopter

What You Need

- ❏ Sheet of 8½-by-11-inch paper
- ❏ Scissors
- ❏ Ruler
- ❏ Pencil
- ❏ Two paper clips

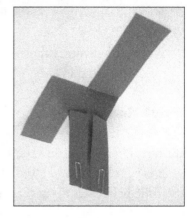

What to Do

Fold the paper in half lengthwise and cut along the fold with scissors. Fold one of the halves in half lengthwise.

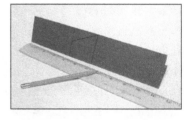

From the base of the paper, measure 4 inches up the length of the folded sheet of paper and draw a line across the width of the paper. From the base of the paper, measure 6 inches up the length of the folded sheet of paper and draw a line across the width of the paper.

On the line drawn at 4 inches, measure 1 inch in from the open edge of the folded paper. Draw a diagonal line from this point to the point where the line drawn at 6 inches touches the open edge of the folded paper.

Cut out the triangle, cutting through both layers of paper.

155

Open the paper and cut the center fold from the top of the paper to the line drawn at 6 inches.

Fold the tabs at the bottom of the paper toward the center and attach two paper clips to the bottom.

Fold the wings in opposite directions along the line drawn at 6 inches.

Hold the helicopter above your head and drop it.

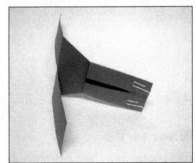

What Happens

The paper helicopter rotates like a real helicopter, and the more paper clips you add, the faster the helicopter rotates.

Why It Works

As the helicopter falls, air rushes out from under the wings in all directions. The air hits the body of the craft, causing it to rotate around a central point. Adding more paper clips increases the weight and reduces the air resistance but increases the amount of air hitting the helicopter wings.

Bizarre Facts

■ Although Russian-born Igor Sikorsky is credited with inventing the first successful single-rotor helicopter in 1939, French inventor Paul Cornu built the first manned helicopter in 1907.

■ The principle of the rotary wing was used some 2,500 years ago by the Chinese for the flying top—a stick with propeller-like blades on top that was spun into the air as a children's toy.

■ The hummingbird, the only bird capable of hovering in the air, beating its wings up to seventy times a second, provided the inspiration for the helicopter.

Plastic Milk

What You Need

- ❏ 1 cup whole milk
- ❏ Small saucepan
- ❏ Measuring spoons
- ❏ White vinegar
- ❏ Clean, empty glass jar
- ❏ Wax paper

What to Do

Pour the milk into the saucepan, add 2 teaspoons of vinegar, and heat, stirring frequently. The milk will boil and then form tiny lumps (curds) in a clear liquid (whey). Slowly pour off the liquid from the pot into the sink. Then spoon the curds into the jar.

Add 1 teaspoon vinegar to the curds, and let the mixture stand for one to two hours. The curds will form a yellowish glob at the bottom of a clear liquid. The glob is actually fat, minerals,

157

and the protein casein. Pour out the liquid, remove the rubbery yellow glob from the jar, wash the glob with water, and knead it until it attains the consistency of dough.

Mold the plastic into any shape you wish, and then place it on the wax paper. Let dry overnight.

What Happens
The casein from the milk hardens into plastic that can be painted with acrylic paints.

Why It Works
The combination of heat and acetic acid precipitates the casein, an ingredient used to make plastic, from the milk.

Bizarre Facts
■ The Japanese developed a low-cost, biodegradable plastic made from shrimp shells by combining *chitin*—an extract from the shells that is also found in human fingernails—with silicon. The resulting "chitosan" is stronger than petroleum-based plastics, decomposes in soil, and acts as a fertilizer.

■ In 1929, the Borden Company purchased the Casein Company of America, the leading manufacturer of glues made from casein, a milk by-product. Three years later, Borden introduced its first nonfood consumer product, Casco Glue.

■ You can make glue from milk by simply adding ⅓ cup of vinegar to 1 cup of milk in a widemouthed jar. When the milk separates into curds and whey, pour off the liquid and wash it away. Add ¼ cup of water and 1 tablespoon of baking soda. When the bubbling stops, you've got glue.

■ Cheese is made from curds. White glue is made from the casein of the curds.

■ Twelve or more cows are called a *flink*.

■ Before scientists discovered how to synthesize plastics from petroleum products, plants and animal fats were used to make natural plastics, which eventually decompose.

■ Biodegradable plastic is made by adding starch to the plastic. Bacteria then feed on the buried plastic.

■ In surgery, stitches are now made using plastics that slowly dissolve in body fluids.

■ The Sanskrit word for *war* means "desire for more cows."

■ There are more plastic flamingos in America than real ones.

■ Cow's milk is 87 percent water.

■ In Arctic regions, people get milk from reindeer.

■ In Peru and Bolivia, people drink llama's milk.

■ Twenty-four percent of all Americans drink milk with dinner.

■ The glue on Israeli postage stamps is certified kosher.

■ Ben and Jerry's sends the waste from making ice cream to local pig farmers to use as feed. Pigs love it, except for one flavor: Mint Oreo.

■ On May 23, 1992, Ashrita Furman of Jamaica, New York, walked 61 miles with a full pint bottle of milk balanced on his head.

Just One Word

In the 1967 movie *The Graduate*, starring Dustin Hoffman, during a party in his parents' home to celebrate his college graduation, Benjamin Braddock is steered into the backyard by a concerned friend of his parents.

MR. McQuire: Ben, I want to say one word to you—just one word.

BENJAMIN: Yes, sir.

MR. McQuire: Are you listening?

BENJAMIN: Yes I am.

MR. McQuire: *(gravely)* Plastics.

159

Play Dough

What You Need

- Large glass bowl
- Food coloring
- 2 cups water
- 2 cups flour
- 1 cup salt
- 1 teaspoon alum
- 1 tablespoon corn oil
- 1 teaspoon 20 Mule Team Borax
- Frying pan
- Wooden spoon
- Cutting board
- Ziploc Storage Bag or airtight container

What to Do

In the bowl, combine the water and fifty drops of food coloring. Then add the flour, salt, alum, corn oil, and borax. Mix well. With adult supervision, cook and stir over medium heat for three

160

minutes (or until the mixture holds together). Turn onto the cutting board and knead to proper consistency. Store in the Ziploc Storage Bag or airtight container.

What Happens
The alum and borax prevent bacteria from turning the thick, colored dough sour.

Why It Works
The patent for Play-Doh puts it this way: "We are unable to state definitely the theory upon which this process operates, because the reactions taking place in the mass are complicated."

Bizarre Facts
▓ In 1956 in Cincinnati, Ohio, brothers Noah W. McVicker and Joseph S. McVicker, employees of Rainbow Crafts, a soap company, invented Play-Doh and received a patent for it in 1965. Kenner acquired Play-Doh, only to be bought out by Hasbro, which transferred Play-Doh to its Playskool division.

Mmmmm, Good!
Kids eat more Play-Doh than crayons, fingerpaint, and white paste combined.

▓ The patent for Play-Doh (U.S. Patent No. 3167440) can be viewed on the internet on the United States Patent and Trademark Office web site at: http://patft.uspto.gov/netahtml/PTO/patimg.htm
▓ After you make Play Dough in an old pan, the pan will be sparkling clean. The combination of flour and salt cleans the pan.
▓ Play-Doh is available in fifty colors.
▓ The Play-Doh boy, pictured on every can of Play-Doh, was created in 1960 and is named Play-Doh Pete.

Plunger Power

What You Need

❑ Two plungers   ❑ An assistant

What To Do

Position the rims of the rubber cups on the ends of the plungers together and push firmly together until the suction cups collapse into each other.

Have an assistant grasp the wooden pole on one plunger while you grasp the second pole. Try to pull the two plungers apart.

What Happens

The rubber plungers remain stuck together, no matter how hard you pull.

Why It Works

Pushing the two rubber cups together forces the air from between the cups and produces a vacuum in the enclosed space. Since no air can get inside the two cups, they remain vacuum-sealed together. They only way to separate the two plungers is to break the seal by prying the plungers apart and allowing air inside.

Bizarre Facts

- A plunger is a large suction cup.

- Trumpet players frequently use the rubber cup of a plunger to muffle the sound emitted from their horn.

- A plunger can be used to pull out a shallow dent in a car door. Simply wet the plunger, place over the dent, and pull out abruptly.

- Before drilling into a ceiling, cut a small hole through the center of the rubber cup of a plunger, and place the cup over the drill bit to catch falling chips of plaster.

- Insert the wooden handle of a plunger into the ground outdoors to create a candleholder.

Take the Plunge

In the 1960s, a hippie in North Miami Beach, Florida, was seen carrying a plunger and wearing a burlap vest emblazoned on the back with the phrase: "Make Love, Don't Plunge into War."

Potato Popper

What You Need

- Knife
- Cutting board
- Several large potatoes
- PVC pipe (¾-inch in diameter and 12-inches in length)
- Wooden dowel (⅝-inch in diameter and 12-inches in length)

What To Do

With adult supervision, use the knife and cutting board to cut the potatoes into slices approximately 1-inch thick.

Press an open end of the PVC pipe down onto one of the potato slices like a cookie cutter to cut out of disk that remains in the tube like a cork. Repeat with the other end of the PVC pipe.

Aiming one end of the PVC pipe into the air, briskly push the wood dowel into the bottom end of the pipe.

What Happens

A potato cork shoots ten to fifteen feet from the pipe, making a funky popping sound.

Why It Works

When you push the wooden dowel into the PVC pipe, it pushes the first potato disk farther into the tube, compressing the air inside the tube. The increased air pressure forces the second potato disk to spring from the tube.

Bizarre Facts

■ The potato originated in the Peruvian and Bolivian Andes mountains, where farmers cultivated it as early as 200 C.E.

■ The Incas invented freeze-dried potatoes. They left potatoes out for several days, allowing them to continually freeze by night and thaw by day, then squeezing out the remaining moisture by hand, and then drying the potatoes in the sun.

■ Spanish conquistadors first introduced the potato to Europe in the late sixteenth century.

■ From 1847 to 1850, potato blight, a fungus disease, swept across Ireland, destroying the country's entire potato crop for four consecutive years. The resulting famine killed more than one million people. Since then, blight-resistant strains of the potato were imported from South America.

■ A statue of Sir Francis Drake in Offenbach, Germany, wrongly proclaims the English explorer "Introducer of the Potato into Europe." There is no evidence that potatoes were aboard his ship, the *Pelican*.

> ## You Say Potato
>
> In 1946, the first toy commercial aired on television. The toy advertised was Mr. Potato Head.

Potato Radio

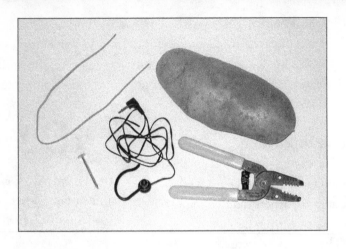

What You Need

- ❏ Wire cutters
- ❏ 12-inch length of 18-gauge copper wire
- ❏ Earphone from a headset
- ❏ One galvanized nail (1-inch long)
- ❏ Potato

What To Do

With the wire cutters, cut the wire in half, and strip 1 inch of insulation off both ends of both pieces of wire. Wrap the end of one wire around the nail, just below the head. Insert the nail into one side of the potato.

Insert one stripped end of the second copper wire into the potato (without letting the wire and the nail touch each other inside the potato).

Cut off the plug at the end of the earphone wire, separate the two wires for 6 inches, and expose the two wires.

166

Attach one of the exposed wires from the earphone to one of the copper wires coming from the potato.

Place the earphone in your ear. Touch the second earphone wire to the free end of the second copper wire.

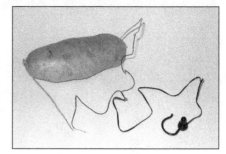

What Happens

You hear static through the earphone.

Why It Works

The citric acid in the potato acts as an electrolyte, conducting an electron flow between the copper in the wire and the iron in the nail, turning the potato into a battery. The resulting electrical current flows in the earphone wires, causing a static sound when the wires touch, when contact is broken, or when you rub the wires against each other.

Bizarre Facts

▣ The leaves of the potato plant are poisonous if eaten.

▣ The eyes in a potato are the indents where sprouts grow. You can grow potatoes by chopping up a potato into chunks with at least two or three eyes, letting the chunks dry in the sun for 24 hours, and then planting them.

▣ Store-bought potatoes are frequently treated with a sprouting inhibitor. If you plant them, they may not grow.

▣ Vincent Van Gogh's 1885 painting *The Potato-Eaters* portrays a family of five peasants gathered around a table, eating potatoes.

▣ In 2010, Peter Glazebrook of Hallam, England, grew the largest recorded potato in history, weighing eight pounds, four ounces.

Psychedelic Milk

What You Need

❑ Paper plate
❑ Milk

❑ Food coloring
❑ Dawn dishwashing liquid

What You Do

Pour the milk into the plate. Near the edge of the milk, add two drops of each color (red, blue, green, yellow) of food coloring. Add one drop of Dawn dishwashing liquid at the center of the milk.

What Happens

The food coloring begins to make wild psychedelic swirls, continuing to dance in the milk for about two minutes. Then the colors mix together and turn muddy gray.

How It Works

At first the drops of food coloring in the milk remain separate because water-based food coloring does not mix with the fat in the milk, but at the same time, the surface water molecules in the milk pull on the puddles of color, spreading them equally in all directions. The dishwashing liquid weakens the pull of the water molecules in the center, causing the stronger water molecules along the rim of the plate to pull the puddles of color toward them. The dishwashing liquid breaks down the drops of fat in the milk, allowing the food coloring and the milk to mix.

Bizarre Facts

■ Abraham Lincoln's mother died from drinking the milk of a cow that had eaten poisonous white snakeroot.

■ The sap of the South American milk tree (*Brosimum utile*) looks, tastes, and can be used just like cow's milk.

■ In 1982, Urbe Blanca, a cow in Cuba, produced 241 pounds of milk in one day—a world record. That's enough milk to provide 120 people with nearly a quart of milk.

■ In 1937, Andy Faust of Collinsville, Oklahoma, milked 88.2 gallons of milk by hand in twelve hours—a world record.

■ In the eighteenth and nineteenth centuries, unscrupulous food manufacturers used colorings to disguise spoiled foods.

■ In 1856, Sir William Henry Perkins discovered the first synthetic dye, derived from coal-tar.

■ In the United States, the first federal regulation concerning food colors was an 1886 act of Congress allowing butter to be colored.

■ Studies show that people judge the quality of food by its color. In fact, the color of a food actually affects a person's perception of its taste, smell, and feel. Researchers have concluded that color even affects a person's ability to identify flavor.

Color My World

In 2004, the average American consumed 15 milligrams of certified food color additives every day.

169

Quicksand

What You Need

- Newspaper
- Measuring cup
- 1¼ cups cornstarch
- 1 cup water
- Large bowl
- Spoon
- 2 tablespoons dry coffee grounds

What to Do

Cover the tabletop with newspaper. Combine the cornstarch and water in the bowl and stir until the mixture looks like a thick and sticky paste. Lightly sprinkle the dry coffee grounds on top of the mixture to give it a dry, even look.

Make a fist and pound on the surface of the quicksand. Then lightly push your fingers down into the mixture.

What Happens

When you hit the surface of the mixture with your fist, the quicksand supports your weight, but when you push your fingers into the mixture, they easily sink to the bottom of the bowl. The coffee grounds make the mixture look deceptively smooth and dry, much like real quicksand.

Why It Works

The cornstarch-water mixture is a *hydrosol*—a solid dispersed in a liquid. When you punch the quicksand, your fist forces the long starch molecules closer together, trapping the water between the starch chains to form a semirigid structure. When the pressure is released, the cornstarch flows again.

Bizarre Facts

▨ Many people have lost their lives sinking into quicksand—a thick body of sand grains mixed with water that appears to be a dry, hard surface. Although it looks as if it can be walked on, quicksand cannot support heavy weight.

▨ Cornstarch is the starch found in corn. All green plants manufacture starch through photosynthesis to serve as a metabolic reserve, but it wasn't until 1824 that Thomas Kingsford developed a technique for separating starch from corn, founding Kingsford's Corn Starch.

▨ Cornstarch is an antidote for iodine poisoning.

▨ Cornstarch makes an excellent substitute for baby powder and talcum powder. Cornstarch is actually more absorbent then talcum powder, but apply it lightly since it does cake more readily.

▨ Cornstarch absorbs excess polish from furniture and cars. After polishing furniture or a car, sprinkle on a little cornstarch and rub with a soft cloth.

Racing Cans

What You Need

❏ Two coffee cans
(11.5 ounces) with
plastic lids
❏ Water
❏ Two pieces of ¾-
inch pinewood,
1-by-3 foot
❏ Two books, both the
same thickness (at
least 1 inch)

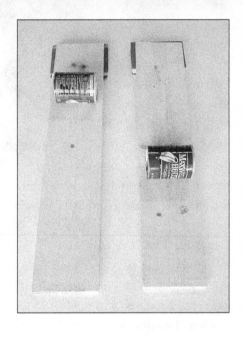

What To Do

Fill the first coffee can
halfway with water and
seal the plastic lid. Fill
the second coffee can to the top with water and seal the plastic
lid.

Place one book on the floor. Prop up the end of one board on
the book to create an incline. Construct a second incline next to
the first.

Place each coffee can on its side at the top of the incline. Release the cans simultaneously and let them roll down the inclines.

What Happens

The filled can rolls faster than the half-filled can and comes to a
stop. The half-filled can rolls to the same spot, then stops and rolls
backward—rolling a greater distance than the full coffee can.

How It Works
The rotational inertia of the half-filled can is greater than the rotational inertia of the full can. The full coffee can slows down due to the friction between the water and the inside of the can. The half-filled can rolls farther because the air inside allows the water to glide inside with less friction, giving the half-filled can greater rotational inertia and making it more difficult to stop.

Bizarre Facts
■ English physicist Sir Isaac Newton first described inertia in 1687 in his first law of motion, which states that a body in motion remains in motion at a constant speed and in the same direction and a body at rest remains at rest unless acted upon by an outside force.

■ Before Sir Isaac Newton died in 1727, his last words were, "I do not know what I may appear to the world; but to myself I seem to have been only like a boy playing on the seashore, and diverting myself in now and then finding a smoother pebble or prettier shell than ordinary, whilst the great ocean of truth lay all undiscovered before me."

Rainbow Machine

What You Need

❏ Clean, empty 1-liter plastic soda
   bottle with cap
❏ Funnel
❏ Measuring cup
❏ 1 cup honey
❏ 1 cup Karo light corn syrup
❏ 1 cup water
❏ 1 cup olive oil
❏ 1 cup rubbing alcohol
❏ Food coloring (red and blue)

What To Do

Using the funnel, slowly pour the
honey into the bottle. When the
honey has settled to the bottom, tilt
the bottle and slowly pour in the corn
syrup, letting it dribble down the in-

side wall of the bottle to form a second layer on top of the honey.
Add five drops of red food coloring to the water and stir well with
the spoon. Tilt the bottle and slowly pour in the colored water,
letting it dribble down the inside wall of the bottle to form a
third layer. Follow the water with the olive oil. Add five drops of
blue food coloring to the alcohol and stir well. Tilt the bottle and
slowly pour the alcohol into the bottle. The layers float on top
of each other without mixing. Secure the cap on the bottle and
vigorously shake the bottle, mixing the contents together. Set the
bottle on a table and observe. Then let sit for eight hours.

What Happens

After being shaken, the layers mix together forming a purple liquid. Eight hours later, the liquid separates into five distinct layers.

Why It Works

Each of the various liquids are immiscible, meaning they do not mix together. Shaking the bottle mixes the immiscible liquids together temporarily into an emulsion, but the emulsion soon separates again. The honey falls to the bottom because it is the densest liquid. The alcohol, being the least dense liquid, rises to the top layer. The food coloring remains mixed in the water because the food coloring is water soluble.

Bizarre Facts

■ An emulsion is a mixture of one liquid evenly dispersed in another liquid—with tiny droplets of the dispersed liquid suspended in the other liquid.

■ Adding an emulsifying agent to an emulsion prevents the immiscible liquids from separating. For instance, a few drops of dishwashing liquid added to a mixture of oil and water keeps the oil suspended in the water.

■ Milk is an emulsion of butterfat in water. The emulsifying agent in milk is the protein casein.

■ Honey never spoils. It crystallizes, but if warmed in a microwave, returns to its liquid state.

■ Corn syrup is made by cooking cornstarch and water under pressure and adding enzymes to the mixture.

Recycled Paper

What You Need

- Newspaper
- Measuring cup
- Clean, empty, large glass jar with lid
- Hot tap water
- Wooden spoon
- Electric blender
- 3 tablespoons cornstarch
- Metal baking pan (larger than 8-by-10 inches)
- Metal screening (8-by-10 inches)
- Scissors
- Markers, crayons, paints, pencils, or pens

What to Do

Cut sheets of newspaper into long, thin strips (or feed the newspaper through a paper shredder) until you have 1½ cups of packed, shredded newspaper.

Put the shredded newspaper into the jar

176

and fill it ¾ full of hot tap water. Screw on the lid and let stand for three hours, shaking the jar occasionally and beating and stirring with the wooden spoon. As the paper absorbs the water, add more hot tap water.

When the mixture becomes pasty and creamy, pour it into the blender—with adult supervision. Dissolve the cornstarch in ½ cup hot tap water, pour into the blender, and blend. Pour the

mixture into the baking pan.

Place the metal screen on top of the mixture in the baking pan, and then gently push it down into the pan until the mixture covers it.

Bring the screen up, place it on a sheet of newspaper, and press it flat with the palm of your hand to squeeze away the water.

Let the screen-backed paper mixture dry in the sun for several hours. When the paper is thoroughly dry, peel it from the screen backing and trim the edges with scissors.

What Happens

The recycled newspaper has the texture of a gray cardboard egg carton and can be decorated with markers, crayons, paints, pencils, or pens.

Why It Works

Newspaper pulp—a blend of sulfite pulp and ground cellulose fibers—when formed into a sheet over a screen, dries into paper again. The cornstarch—a sizing material—is added to the mixture to give the paper a smooth surface and prevent too much ink absorption. Any discarded paper—paper bags, computer paper, junk mail—can be made into pulp.

Bizarre Facts

■ The first paper, invented in China in 105 C.E. by Ts'ai Lun, the Emperor Ho-Ti's minister of public works, was made from the inner bark of the mulberry tree, fishnets, old rags, and waste hemp.

■ For hundreds of years, paper was made by hand from the pulp of rags. Rag pulp is still used today to make high-quality bond paper.

■ Toilet paper and facial tissues are made from wood pulp treated with plant resins to make it absorbent.

■ The average American uses 748 pounds of paper and paperboard every year.

■ U.S. currency paper is 75 percent cotton and 25 percent linen.

■ Before the advent of paper, most documents were written on parchment (made from the skin of sheep or goats) or vellum (made from the skin of calves). A single book three hundred pages long would require the skins of an estimated eighteen sheep.

■ The watermark was discovered by accident. In 1282, a small piece of wire caught in the paper press being used at the Fabrino Paper Mill made a line in the finished paper that could be seen by holding the paper up to the light. The papermakers realized a design made from wire would create a decorative watermark, which could also be used on banknotes to thwart counterfeiters.

Paper Tiger

Paper can be made inexpensively from hemp. However, in 1937, cotton growers, fearing competition from hemp growers, lobbied against marijuana (the dried leaves of the hemp plant) to make hemp illegal. In 1999, Governor Jesse "The Body" Ventura signed legislation making hemp farming legal in Minnesota.

Rising Golf Ball

What You Need

❑ Golf ball
❑ Clean, empty glass jar and lid
❑ Uncooked rice

What to Do

Place the golf ball in the bottom of the jar, and then fill the jar with the uncooked rice, stopping 1½

inches from the top of the jar. Close the lid. Shake the jar back and forth vigorously (not up and down).

What Happens

The golf ball rises to the surface.

Why It Works

No two pieces of matter can occupy the same space at the same time. As the jar is shaken, the grains of rice move closer together, settling in the jar and pushing the ball upward.

Bizarre Facts

▪ More people play golf than any other outdoor sport.
▪ As of January 1, 2012, there were 15,751 golf facilities (a complex containing at least one golf course) in the United States.

■ In 1457, the parliament of King James II of Scotland banned golf and soccer because the popularity of the two sports threatened the practice of archery for national defense. The ban lasted until 1502, when England and Scotland signed a treaty of "perpetual peace."

■ The "birdie"—the term for scoring one stroke under par on a hole—probably got its name from the "feathery"—the name for the original golf ball used until 1848, which was made from leather and stuffed with feathers.

■ At the center of most golf balls is a solid rubber core or a hollow rubber core filled with liquid, usually salt water and corn syrup.

■ Golf balls always spin backward when struck, which, if the balls were smooth, would create higher air pressure above the ball, preventing it from traveling more than seventy yards. Dimples in the ball carry air upward over the top, creating low air pressure over the ball, allowing it to be driven up to three hundred yards.

■ The original Uncle Ben was an African-American rice farmer known to rice millers in and around Houston in the 1940s for consistently delivering the highest quality rice for milling. Uncle Ben harvested his rice with such care that he purportedly received several honors for full-kernel yields and quality. Legend holds that other rice growers proudly claimed their rice was "as good as Uncle Ben's." Unfortunately, further details of Uncle Ben's life (including his last name) were lost to history.

■ Frank Brown, a maître d' in a Houston restaurant, posed for the portrait of Uncle Ben.

■ Rice is thrown at weddings as a symbol of fertility.

■ The world's leading producer of rice is China. The world's leading exporter of rice is Thailand. The world's leading importer of rice is the Philippines.

■ Rice, grown on more than 11 percent of the earth's farmable surface, is the mainstay for more than 3 billion people worldwide.

Rock Candy

What You Need

- Scissors
- Cotton string
- Clean, empty glass jar
- Water
- Spoon

- 2½ cups sugar
- Measuring cup
- Nail
- Pencil

What to Do

Cut a piece of string approximately 1 inch longer than the height of the glass jar. Saturate the string with water, squeeze out the excess liquid, and roll the string in 1 teaspoon of sugar. Let the sugarcoated string dry overnight.

Carefully fill the jar with 1 cup boiling hot water and slowly

181

add 2 cups of sugar, stirring well until the sugar dissolves. Continue adding one teaspoon of sugar at a time, stirring well, until no more sugar dissolves into the saturated solution.

Attach the nail to one end of the sugarcoated string and the pencil to the other end of the string so when you rest the pencil across the mouth of the jar, the nail hangs down into the thick sugar water without touching the bottom of the jar. Place the jar in a warm place and let it stand undisturbed for up to seven days.

What Happens
The water evaporates and rocky sugar crystals form on the string.

Why It Works
As the supersaturated solution cools, the dissolved sugar reemerges to form cube-shaped crystals.

Bizarre Facts
■ The average American consumes 68 pounds of sugar every year. That's 22 teaspoons of sugar every day.
■ *Sugar* is slang for "money."
■ During World War II, GIs called a letter from one's sweetheart a "sugar report."
■ Addressing the Canadian Senate and House of Commons in 1941, Winston Churchill said, "We have not journeyed all this way across the centuries, across the oceans, across the mountains, across the prairies, because we are made of sugar candy."

Sugar, Sugar
The hit song "Sugar, Sugar," by the Archies, was the number one bestselling single in 1969, topping the Beatles' "Get Back" and the Rolling Stones' "Honky Tonk Woman."

Rubber Band Ball

What You Need

❑ Bag of colored #64 rubber bands

What To Do

Tie a thick rubber band in a double knot. Wrap other bands around the knotted rubber band. The ball will look awkward until you achieve the size of a golf ball. Continue wrapping rubber bands around the ball until you achieve the size of a baseball. To maintain a spherical shape, place rubber bands over flat or bare spots. On a hard surface, bounce the ball.

What Happens

The ball will bounce like a rubber ball.

Why It Works

Rubber bands form the circumference of a circle, and a sphere is a circle spun around an axis. The rubber bands are made from rubber, and thus the resulting sphere is a rubber ball that bounces.

Bizarre Facts

■ In 1839, Charles Goodyear invented the vulcanization process for rubber after a long and courageous search that bordered on obsession. He died penniless, but today Goodyear tires are found on millions of cars.

■ In 1998, the *Guinness Book of Records* acknowledged that John Bain of Wilmington, Delaware, single-handedly made the world's largest rubber band ball while working in the mail room of a law firm. The ball weighed 2,008 pounds with a circumference of 13 feet, 8½ inches, and was composed of 350,000 rubber bands. You can see pictures of Bain's rubber band ball on his web site: www. recordball.com.

Rubber Chicken Bone

What You Need

❏ Uncooked chicken bone or wishbone
❏ Clean, empty glass jar with lid
❏ White vinegar

What to Do

Clean the bone thoroughly and let dry overnight. Place the bone in the jar and add enough vinegar to cover the bone. Secure the lid and let the jar stand undisturbed for seven days.

What Happens

The bone becomes soft and rubbery. It can be twisted, and, in some cases, tied in a knot.

Why It Works

The vinegar (acetic acid) dissolves the calcium from the bone, leaving it soft and bendable.

Bizarre Facts

▪ The Etruscans (the people of ancient Northern Italy) originated the superstition of having two people each make a wish and tug on the opposite ends of the dried V-shaped clavicle of a fowl.

The person who breaks off the larger piece of the "wishbone" allegedly has his or her wish come true. Etymologists claim this custom gave birth to the expression "get a lucky break."

■ *Boneyard* is slang for "cemetery."

■ Before double murderer Robert Harris was executed in California in 1992, his last meal included a bucket of Kentucky Fried Chicken.

■ Colonel Sanders told the *New York Times* that the eleven herbs and spices in Kentucky Fried Chicken "stand on everybody's shelf." In reality, there are only four ingredients, according to author William Poundstone, none of which are herbs: flour, salt, monosodium glutamate, and black pepper.

The Funky Chicken

Colonel Harland Sanders, who, at the age of 61, founded Kentucky Fried Chicken, talked his wife into hiring his mistress as their live-in housekeeper, according to the Colonel's daughter, Margaret Sanders, in her book, *The Colonel's Secret: Eleven Secret Herbs and a Spicy Daughter*. Sanders later divorced his wife, married his mistress, and took both women with him to Washington, D.C., to attend a presidential inauguration.

Seesawing Candle

What You Need
❑ Knife
❑ Candle
❑ Needle
❑ Wax paper
❑ Two drinking glasses
❑ Fire extinguisher
❑ Matches

What to Do
With adult supervision, use the knife to carefully carve away the tallow or wax at the bottom end of the candle to expose the wick. Carefully push the needle through the center of the candle.

Place a piece of wax paper on the tabletop, set the drinking glasses on the wax paper, and rest the needle across the rims of each glass so the candle is between the glasses. With a fire extinguisher on hand, light both ends of the candle and observe.

What Happens
The candle starts rocking like a seesaw and continues as long as both ends stay lit.

Why It Works
As Sir Isaac Newton's third law of motion states: For every action there is an opposite and equal reaction. When the tallow or wax drips off each end of the candle, it delivers an upward recoil.

187

Bizarre Facts

▦ During Roman times, starving soldiers ate their candle rations, which were made from tallow—a colorless, tasteless extract of animal or vegetable fat.

▦ British lighthouse keepers, isolated for months, often ate their tallow candles.

▦ The charred end of the wick of a tallow candle had to be "snuffed"—snipped off without extinguishing the flame—every half hour, otherwise the candle provided only a fraction of its potential light and the low-burning flame burned only 5 percent of the tallow, melting the remaining tallow.

▦ Because snuffing a tallow candle was difficult to do without extinguishing the flame, the word *snuff* came to mean "extinguish."

▦ Beeswax oozes from small pores in the abdomen of a worker bee and forms tiny white flakes on the outside of its abdomen. Using its legs, the bee picks off these flakes and moves them to its jaws. The bee then chews the wax to the proper consistency to build the honeycomb.

▦ Native Americans in the Pacific Northwest used dried candlefish, a saltwater fish about eight inches long, as candles.

▦ Spermaceti, a waxy material obtained from the enormous head of the sperm whale (making up a third of its body), was once used to make candles.

▦ The world's largest candle on record, displayed at the 1897 Stockholm Exhibition, was eighty feet high and 8½ feet in diameter.

▦ The world's largest needle, measuring 6 feet, 1 inch long, used for stitching on mattress buttons lengthwise, can be visited at the National Needle Museum in Forge Mill, Great Britain.

Self-Inflating Ball

What You Need

- Safety goggles
- Electric drill with ¼-inch bit and ¾-inch bit
- Cork
- Eyedropper
- Beach ball (1 foot in diameter)
- Scissors
- Black electrical tape
- Bucket
- Wooden mixing spoon
- 1 cup sugar
- 2 tablespoons molasses
- Packet of yeast (¼ ounce)
- 3 quarts water

- Funnel
- Clean, empty, 1-gallon Gatorade bottle

What To Do

With adult supervision and wearing safety goggles, drill a ¼-inch hole through the middle of the cork. Insert the glass tube from the

189

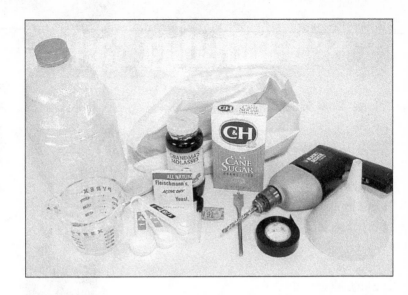

eyedropper through the hole in the cork so the tip protrudes from the top of the cork.

Insert the tip of the eyedropper into the nozzle on the uninflated beach ball and secure in place with a piece of electrical tape.

Drill a ¾-inch hole through the center of the lid of the Gatorade bottle. Insert the cork firmly in the hole in the lid.

In the bucket and using the wooden spoon, mix the sugar, molasses, yeast, and water until dissolved. Using the funnel, pour the mixture into the Gatorade bottle. Seal the prepared lid (complete with the cork and attached beach ball) onto the bottle. Set in a warm place for several days.

What Happens

The beach ball inflates with carbon dioxide.

How It Works

The yeast is a fungus that grows in the mixture, converting the sugar into alcohol and carbon dioxide gas, which fills the ball.

Bizarre Facts

■ The resulting liquid in this experiment is a fermented mixture of crude ethyl alcohol that can be made into rum through a complicated distilling process.

■ The yeast used to ferment alcoholic beverages is actually a fungus. Members of the same fungus family (Asomycetes) produce the antibiotics penicillin and streptomycin.

■ In 1876, French scientist Louis Pasteur first reported that yeasts were living cells.

■ Bakers add yeast to dough to make it rise. In bread-making, the yeast converts the sugar in the flour into alcohol and carbon dioxide. Bubbles of gas get trapped by the gluten in the dough, and as the gas expands, the bread rises. Baking destroys the yeast and causes

Murder by Breathing

If you are locked in a sealed room, you will die of carbon dioxide poisoning—before all the oxygen is depleted from the room.

the alcohol to evaporate from the bread. Before yeast began being produced commercially in the 1880s, bread-makers prepared dough and left it uncovered so that airborne yeast plants landed on it and began the fermentation process.

■ Model ships in narrow-necked bottles are made by inserting the finished model into the bottle with its hinged masts and sails lying down, then pulling a thread to draw the rigging upright.

■ Liaoning, China, is home to the largest bottle in the world. Measuring 15 feet tall, the bottle holds 488 gallons of liquid.

Silver Egg

What You Need

- Candle
- Matches
- Saucer
- Tweezers
- Piece of eggshell
- Glass of water

What To Do

With adult supervision, melt the bottom of the candle and secure it upright in the center of the saucer. Light the candle.

Using the tweezers, hold the eggshell over the flame until it is smoked black. Then submerge the eggshell in the glass of water.

What Happens

The egg shell appears to be silver.

How It Works

The flame deposits lampblack and small amounts of cracked paraffin on the eggshell. This coating is a hydrophobic mixture,

meaning it repels water. Thousands of tiny air bubbles accompany the submerged eggshell, preventing the black coating from getting wet. Light reflects from these air molecules—making the egg shell appear silver.

Bizarre Facts

■ In psychology, the word *hydrophobic* means "fear of water," but in chemistry the same word means "having no affinity for water."

■ No word in the English language rhymes with silver.

■ In L. Frank Baum's classic children's book, *The Wonderful Wizard of Oz*, Dorothy wears silver shoes. Hollywood screenwriter Noel Langley changed them to ruby slippers in the script for MGM's classic 1939 movie, *The Wizard of Oz*.

■ In 1876, Nell Saunders, winner of the first United States women's boxing match, received a silver butter dish as a prize.

■ More silver is mined in Mexico than any other country, followed by Peru and China.

■ The sixteenth-century astronomer Tycho Brahe, having lost his nose in a duel with one of his students, wore an artificial nose made from silver.

Slinky Race

What You Need
☐ Traditional metal Slinky
☐ Metal Slinky Jr.
☐ Staircase

What To Do
Place a Slinky and a Slinky Junior at the top of a staircase. Simultaneously flip the top coil of each Slinky to the next lower step and let go.

What Happens
The smaller Slinky beats the larger Slinky down the stairs.

Why It Works
A Slinky sitting at the top of a staircase has *potential energy*. When a force is applied to the Slinky to make it start down the stairs, the Slinky is affected by gravity. The potential energy becomes *kinetic energy*, and the Slinky flips coil over coil down the stairs. As the Slinky flips down the steps, the energy is transferred along its length in a compressional wave (resembling a sound wave).

How fast a Slinky walks down steps depends how quickly a longitudinal wave travels through the Slinky coil, which depends on the tension and mass of the coil. The tighter the coil, the faster the longitudinal wave travels through the Slinky. The smaller the mass of the Slinky, the tighter the tension in the coil. The wave moves faster through the smaller Slinky, making it travel quicker.

Bizarre Facts

■ The Slinky helps scientists understand the supercoiling of DNA molecules. Slinky and Shear Slinky, two computer graphics programs developed at the University of Maryland, use a Slinky model to approximate the double helix coiling of DNA molecules.

■ In 1985, Space Shuttle astronaut Jeffrey Hoffman became the first person to play with a Slinky in orbit around the earth.

■ If you hold one end of a Slinky and whirl it around your head, it swings out from you. This is caused by centrifugal force. The faster you make the Slinky go around, the longer the centrifugal force stretches out the Slinky and raises it from the ground.

■ Physics teachers use Slinkys to teach students about the properties of waves.

■ A Slinky can be seen in the movies *Hairspray* (1988), *Other People's Money* (1991), *Demolition Man* (1993), *The Inkwell* (1994), and *Ace Ventura 2: When Nature Calls* (1995).

■ In 1999, the United States Postal Service introduced the world's first Slinky stamp.

■ Pecan harvesters have used Slinkys to help collect pecans.

A Slinky Mystery

Standing on a ladder, hold a Slinky by one end and let the other end hang down without touching the ground. When you drop the Slinky will the bottom end spring upward? Or will the bottom end fall to the ground first? Or will the bottom end stay where it is until the entire Slinky has collapsed and then fall to the ground?

Smoke Bomb

What You Need

- ¼ cup sugar
- 3 ounces saltpeter (from a drugstore)
- Bowl
- Saucepan
- Spoon
- Paper cup
- Wooden matches
- Clean, empty tin can
- Fire extinguisher

What to Do

With adult supervision, combine the sugar and saltpeter in the bowl, and then heat in the saucepan over very low heat, stirring constantly, until the mixture melts into a plastic substance that resembles caramel. Remove from the heat, pour into the paper cup, embed a dozen wooden matches, heads up, into the hardening substance, and let cool.

When the substance cools and hardens, peel off the paper cup, place the smoke bomb inside the tin can or on a concrete surface (away from flammable objects or areas). With a fire extinguisher on hand, light the match heads, and stand back. (Or you can simply leave the paper cup in place and light it on fire.)

What Happens
The smoke bomb produces a cloud of thick white smoke.

Why It Works
The melted sugar becomes a candy that encapsulates the saltpeter, making it easier to ignite.

Bizarre Facts
■ In ancient Chinese kitchens, saltpeter was commonly used as a preserving and pickling salt.

■ Saltpeter is one of the three original ingredients in fireworks.

■ Smoke is simply small particles made airborne by the buoyancy of the hot gases produced by combustion.

■ Along the Great Wall of China, guards sent smoke signals made with wolf dung, because the smoke hung in the air for a long time.

■ In the 1999 movie *October Sky*, the rocket boys mix saltpeter and sugar to make rocket fuel.

The Joys of Saltpeter
Saltpeter, known to chemists as potassium nitrate, is commonly believed to reduce a man's primitive urges. It doesn't.

Smoking Fingertips

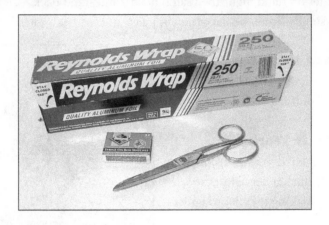

What You Need

❑ Scissors
❑ Box of wooden matches
❑ Aluminum foil

What To Do

With scissors, cut the striking surface from one side of the box of matches. Peal off the extra layer of cardboard from the striker. Fold the striker in half along the length toward the striking surface. Place the folded striker face down on a piece of aluminum foil, and with adult supervision, use a lit match to set the striking surface on fire, leaving an orange rust on the foil.

Discard the striker, use the tip of your index finger to wipe up the orange chemical from the foil, and then rub your index finger and thumb together briskly. Afterward, wash your hands.

What Happens

White smoke rises from your fingertips.

How It Works

The striking surface on a box of wooden matches contains red phosphorus, which is deposited on the aluminum foil after burning the strip. The warmth from the friction created by rubbing your fingers together causes the red phosphorus to bond with the oxygen in the air to form phosphorous oxide—a white vapor.

Bizarre Facts

▓ The fingerprints of koala bears are almost indistinguishable from the fingerprints of human beings.

▓ During World War II, the United States Fleet Commander in the Pacific Theater (Admiral Chester Nimitz) and the Japanese Fleet Commander (Admiral Isoroku Yamamoto) had each lost fingers from their left hands as the result of accidents while they were younger officers aboard ships.

▓ Actor James Doohan, best remembered as Lt. Commander Montgomery Scott on the television series *Star Trek*, lost his right middle finger due to bullet wounds received on D-Day.

▓ Grateful Dead guitarist Jerry Garcia was missing the top two-thirds of the middle finger on his right hand, due to an accident with an axe at age four.

▓ The white, crescent shaped part of your fingernail is called the lunula (from the word *luna*, meaning "moon").

▓ Working in pressurized space suits causes the fingerprints on the fingertips of the Space Shuttle astronauts to be rubbed away.

▓ On the popular television comedy variety series *Rowan & Martin's Laugh-In*, the cast routinely bestowed The Flying Fickle Finger of Fate Award.

Sonic Blaster

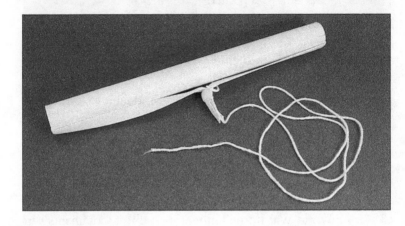

What You Need

- PVC pipe (3¼ inch in diameter and 12 inches in length)
- 20-inch length of elastic
- 5-foot length of string
- Bicycle helmet
- Safety goggles

What To Do

Thread one end of the elastic through the PVC tube and tie the two ends together securely to form a loop. Tie one end of the string around the elastic at one of the tube openings.

Wearing the bicycle helmet and safety goggles, hold the free end of the string and whirl the PVC tube around your head, like a lasso.

What Happens

The PVC tube makes a loud roaring sound as it flies through the air.

How It Works

As the tube spins around, air enters the tube, causing the elastic band to vibrate rapidly. This sound is carried out by the air leaving the tube.

Bizarre Facts

■ The longer the length of string, the louder the resulting sound.

■ In the 1986 movie *Crocodile Dundee*, actor Paul Hogan makes a similar sonic blaster using a rock and a long piece of twine.

■ The sound of knuckles cracking is caused by imploding synovial fluid, the liquid that keeps the joints moist and lubricated.

■ When a bullwhip is cracked, the tip moves so fast that it actually breaks the sound barrier, creating a tiny sonic boom.

■ Cats can hear ultrasound.

■ Novelist William Faulkner titled his novel *The Sound and the Fury* after a line from William Shakespeare's play *Macbeth*: "Life's but a walking shadow, a poor player that struts and frets his hour upon the stage, and then is heard no more; it is a tale told by an idiot, full of sound and fury, signifying nothing."

■ One of the most popular songs recorded by the musical duo Paul Simon and Art Garfunkle is "The Sound of Silence."

■ Any sound louder than 120 decibels is painful to the human ear. Racing cars emit 125 decibels. Rock concerts typically reach 130 decibels.

■ Sound travels one mile through air in approximately five seconds. Sound travels one mile through water in roughly one second.

Sounds Strange

When the pilot of a plane breaks the sound barrier, the pilot does not hear the sonic boom heard by people on the ground (because the pilot is traveling faster than the speed of sound).

Sparkling Life Savers

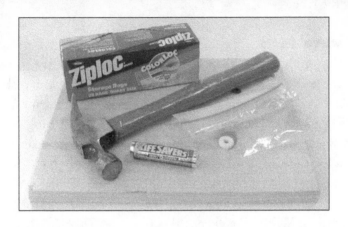

What You Need

❏ Wint-O-Green Life Savers
❏ Ziploc Storage Bag
❏ Wooden block
❏ Hammer

What to Do

Place one Wint-O-Green Life Saver in the Ziploc Storage Bag. Seal the bag and place it on the wooden block. In a dark room or closet, hold the hammer above the Life Saver. Look directly at the Life Saver as you smash it with the hammer.

What Happens

A quick burst of bluish-green light flashes the moment the wintergreen candy is crushed.

Why It Works

Crushing a crystalline substance, in this case the sugar crystals in the Life Savers, produces ultraviolet radiation, which causes

202

the molecules in the synthetic wintergreen flavoring in the candy—methyl salicylate—to fluoresce. This phenomenon is called *triboluminescence*. (Wint-O-Green Life Savers do not always spark consistently. Moisture sometimes absorbs the energy needed to produce sparkling.)

Bizarre Facts

■ In 1912, Cleveland-based chocolate maker Clarence A. Crane developed the idea for white-circle mints, had a pharmaceutical manufacturer produce them on his pill machine, and named them Life Savers because they resembled the flotation device.

■ Advertisements for Life Savers have included the corny puns "hole-some," "enjoy-mint," "refresh-mint," and "content-mint."

■ In 1970, the Dow Chemical Company introduced the Ziploc Storage Bag with its patented tongue-in-groove "Gripper Zipper," providing a virtually airtight, watertight seal that revolutionized plastic bags. "Ziploc" is a clever hybrid of the words *zipper* and *lock*—a mnemonic device to remind consumers that the bags zip shut and lock tight.

Deadly Life Savers?

In a letter published in the *New England Journal* of Medicine, two Illinois physicians, Dr. Howard Edward, Jr., and Dr. Donald Edward, warned that biting a Wint-O-Green Life Saver while in an oxygen tent, operating room, or space capsule could be life-threatening. The *Journal* declared Wint-O-Green Life Savers safe for oxygen tents and gas stations.

Spinning CD

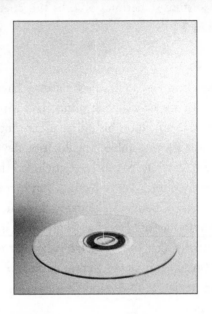

What You Need

❑ 3-foot length of dental floss
❑ Toothpick

❑ Old compact disc
❑ Scotch Tape

What To Do

Tie one end of the piece of string to the middle of the toothpick.
Insert the toothpick through the hole in the compact disc so that
when you hold the other end of the string, the disc rests on the
toothpick and the knot in the dental floss is in the center of the
hole. Use two pieces of Scotch Tape to secure the ends of the
toothpick to the compact disc.

Try to swing the compact disc back and forth like a pendulum,
keeping it level without letting it flop around. Stop. Use your in-

dex finger to spin the compact disc on the string (as if tossing a Frisbee). Now try swinging it back and forth like a pendulum.

What Happens

As you swing the spinning compact disc back and forth, the disc remains level without any effort.

How It Works

The spinning compact disc is a gyroscope, a simple machine that seems to defy the laws of gravity. A spinning body tends to spin on the same plane on which it began spinning, unless an outside force moves it from that original plane. This ability of a spinning body to always spin on the same plane is called *gyroscopic inertia.*

Bizarre Facts

■ Gyroscopes guide airplanes and rockets, control stabilizers on ships, and can be used as a reliable navigational aid.

■ The spinning wheels of a moving bicycle turn the vehicle into a gyroscope, allowing the rider to stay balanced with minimal effort.

■ A spinning gyroscope is unaffected by the earth's gravity.

■ A spinning gyroscope holds its original position in space while the earth turns under it. If you point the axle of a spinning gyroscope at the sun, the axle appears to follow the sun as it crosses the sky.

■ The first gyroscope in recorded history was built by German scientist G. C. Bohnenberger.

■ In 1852, French physicist Jean Foucault built a gyroscope to demonstrate that the earth rotates on its axis. Foucault also named the device after the fact that he used it to view the revolution of the earth, combining the Greek word *gyros* (meaning "revolution") with the Greek word *skope*in (meaning "to view").

Steel Wool Sparkler

What You Need
❑ Steel wool pad
❑ 9-volt battery

❑ Baking pan
❑ Fire extinguisher

What to Do
With adult supervision, pull the steel wool pad apart until it is the size of a tennis ball. Place the steel wool pad in the baking pan. With a fire extinguisher on hand, touch the ends of the battery to the steel wool.

What Happens
The sparks from the battery cause the steel wool to catch on fire, and the iron filings from the steel wool flash like a sparkler.

Why It Works
The threads of iron in the steel wool, surrounded by more oxygen than in a solid block of iron, combust easily.

Bizarre Facts

■ In 1917, Edwin Cox, a struggling door-to-door aluminum cookware salesman in San Francisco, developed in his kitchen a steel wool scouring pad caked with dried soap as a gift to house-wives to get himself invited inside their homes to demonstrate his wares and boost sales. A few months later, demand for the soap-encrusted pads snowballed, and Cox quit the aluminum cookware business and went to work for himself.

■ Mrs. Edwin Cox, the inventor's wife, named the soap pads S.O.S., for "Save Our Saucepans," convinced that she had clev-erly adapted the Morse code international distress signal for "Save Our Ships." In fact, the distress signal S.O.S. doesn't stand for anything. It's simply a combination of three letters represented by three identical marks (the S is three dots, the O is three dashes). The period after the last S was deleted from the brand name to obtain a trademark for what would otherwise be an international distress symbol.

■ *Brillo*, the brand name of a competing steel wool scouring pad, is a derogatory name for a person with tight curly hair.

Be Prepared

The Boy Scout manual now instructs all Boy Scouts to start a fire with a steel wool pad and a 9-volt battery rather than the traditional method of starting a fire with flint and steel.

Submarine

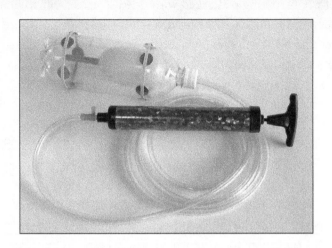

What You Need

- Clean, empty, 1-liter plastic soda bottle with cap
- X-acto knife
- Three rubber bands
- Ten nickels
- Safety goggles
- Electric drill with ¼-inch bit
- 6-foot length of flexible plastic tube (¼ inch in diameter)
- Balloon
- Basketball hand pump
- Black electrical tape

What To Do

With adult supervision, cut five square 1-inch windows along the length of the bottle with the X-acto knife.

Put one rubber band around the top of the bottle and the second rubber band around the bottom of the bottle. Put five nickels under each rubber band, equidistantly surrounding the bottle.

Wearing safety goggles, drill a ¼-inch hole in the middle of

 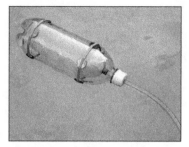

the cap. Fit one end of the tubing snugly through the cap so that it extends 4 inches through the inside of the cap.

Blow up the balloon and then let the air out, so you have stretched the rubber. Put the mouth of the balloon around the end of the 4-inch length of tube and secure it in place with a rubber band. Insert the balloon on the end of the tube inside the bottle, and screw on the cap securely.

Holding the free end of the tube, place the submarine in a swimming pool or a bathtub filled with water. Insert the nozzle of the basketball hand pump into the free end of the tube, secure in place with electrical tape, and inflate the balloon.

What Happens

When you put the submarine in the water, the water enters the holes, filling the bottle with water and causing the bottle to sink. When you pump air into the tube, the balloon inflates, expands, and pushes the water out of the bottle, causing the submarine to float to the surface.

Why It Works

The water-filled submarine is negatively buoyant—or denser than the water (due to the weight of the nickels)—and sinks in the water. When the submarine is filled with air, it becomes buoyant—or less dense than the water—and floats to the surface.

Bizarre Facts

■ A submarine dives by achieving neutral or negative buoyancy by using ballast tanks that can be filled with either water, air, or a combination of both to adjust the ship's buoyancy. To make the submarine submerge, the crew lets water into the ballast tanks to make the ship heavier. To make the submarine rise, the crew uses compressed air to push the water out of the tanks, reducing the ship's weight. A propeller moves the submarine forward.

■ In his 1870 science-fiction novel *Twenty Thousand Leagues Under the Sea*, French author Jules Verne tells the tale of Captain Nemo, a mad sea captain who cruises beneath the oceans in his submarine, the *Nautilus*.

■ During World War II, some forty thousand Germans served aboard Nazi submarines. Approximately ten thousand survived.

■ On the popular 1960s television series *Voyage to the Bottom of the Sea*, Commander Lee Crane, played by actor David Hedison, captained the *Seaview*, a glass-nosed atomic submarine, through weekly adventures. The show was based on the 1961 movie of the same name, starring Walter Pidgeon, Joan Fontaine, Barbara Eden, Peter Lorre, and Frankie Avalon.

■ In the 1968 animated movie *Yellow Submarine*, the Beatles travel aboard a yellow submarine through the Sea of Holes and the Sea of Time to save Pepperland from the Blue Meanies.

■ The 1981 movie *Das Boot* chronicles the claustrophobic conditions aboard a German submarine during World War II.

Dive! Dive! Dive!

In 1620, the first known submarine, the *Drebbel*, was built for King James I of England. It navigated the Thames River and moved by use of oars that extended through sealed oarlocks.

Swinging Cups

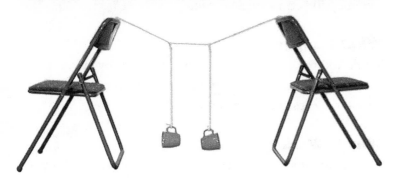

What You Need

- Two chairs
- Thin rope
- Ruler
- Scissors
- Two coffee cups

What to Do

Place the chairs back to back 3 feet apart. Tie a piece of the rope from the top of one chair to the top of the other.

Measure and cut two pieces of thin rope 24 inches long and tie a coffee cup to the end of each rope. Tie the free ends of the ropes to the rope between the chairs, letting the coffee cups hang 20 inches apart from each other and equidistant from the chairs. Move the chairs so that the center of the horizontal rope sags about 4 inches below the spots where the cord is tied to the chair backs.

Start one of the cups swinging at right angles to the horizontal cord like a pendulum.

Tell your audience to watch closely. Insist that by using your hypnotic energy, you will make the first cup stop swinging and the second cup start swinging. Wiggle your fingers toward the

211

cups. The first cup slows down until it stops completely while the second cup begins swinging. A moment later, wiggle your fingers toward the cups again. The second cup will stop swinging and the first one will begin swinging again.

What Happens

While you pretend to be controlling the cups hypnotically, the motion continues being transferred from one cup to the other as long as the cups keep swinging.

Why It Works

Pendulums attached to a central line transfer energy back and forth between themselves through the connecting line.

Bizarre Facts

■ Galileo Galilei first demonstrated this effect in 1583.

■ At age twenty, Galileo discovered the law of the pendulum by timing the swings of a bronze lamp that hangs from the ceiling in the Duomo in Pisa, Italy. He observed that each swing took the same time, whether the arc was large or small.

■ Edgar Allen Poe, author of the 1843 short story "The Pit and the Pendulum," married his cousin Virginia Clemm when she was not quite fourteen years old.

World's Largest Yo-Yo

On July 6, 2010, Jerry Havill's Team Problem Solving course at Bay de Noc Community College in Escanaba, Michigan, launched a yo-yo measuring 11.5 feet in diameter and weighing 1,625 pounds from a height of 100 feet from a crane. It "yo-yoed" thirty times. The word yo-yo means "come-come" in Filipino, and the toy purportedly originated in the Philippines as a weapon.

Switching Switches

What You Need

- Light-bulb base
- Plastic project enclosure box (6-by-3-by-2 inches)
- Safety goggles
- Electric drill with ⅜-inch bit and ¹⁄₁₆-inch bit
- Precision screwdrivers
- Needle-nose pliers
- Three push-on/push-off switches
- Krazy Glue
- AA battery holder
- Wire cutters
- 2-foot length of 22-gauge electrical wire
- Soldering iron
- Solder
- AA battery
- 2-volt light bulb

What to Do

Mount the light-bulb base on the cover panel of the project enclosure box. With adult supervision and wearing safety goggles, use the drill with the ⅜-inch bit to mount the three push-on/push-off switches under the light bulb base. Using Krazy Glue, adhere the battery holder to the underside of the cover panel. Wire

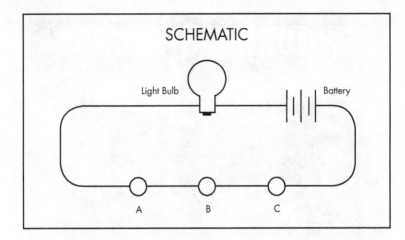

SCHEMATIC

Light Bulb

Battery

A   B   C

the switches in series (as illustrated in the schematic diagram) to the battery holder and light-bulb base.

Wearing safety goggles and with an adult watching, solder the connections. Insert the battery and light bulb. Screw the cover panel into place on the project enclosure box.

Press all the buttons so the bulb lights. Push button A several times to show that it controls the bulb. Leave the switch off. Push button B an odd number of times to show that it does not affect the bulb. Then push button C an odd number of times to show that it does not affect the bulb. This leaves all the buttons turned off.

Ask a volunteer to push whatever button he thinks will turn on the bulb. He naturally pushes button A, but the bulb does not light. Give him a second chance. He pushes button B or C, but again the bulb does not light. You push the remaining button, which miraculously lights the bulb.

Repeat the trick by pushing the last button several times to show it does indeed light the bulb, and then leave it off. Push the other two buttons an odd number of times to show that they do not light the bulb. Once again, ask the volunteer to push whatever button he thinks will turn on the bulb.

What Happens

No matter how many times you repeat the trick, the volunteer continues failing on his first two tries. The third button selected always lights the bulb. Most people assume there is a perplexing circuit pattern under the board.

Why It Works

The switches and the bulb are joined in series to the battery.

All the switches are turned off when the volunteer is asked to push a button. His first choice turns on the first button, his second choice turns on the second button, and the third button (which you push) completes the circuit and lights the bulb.

Bizarre Facts

■ The trick of making a pea vanish under one of three thimbles originated in Egypt some five thousand years ago using cups and balls. The trick is done with sleight of hand. The conjurer lets the pea roll out from under the cup, quickly catches it in the hand holding the cup, and then drops it into another thimble.

■ Three Card Monte—a card trick in which a gullible spectator is asked to wager money on which one of the three cards lying face down on a table is the queen—is rigged by a fake throw. The operator, previously throwing the bottom card from his right hand before the top card, unexpectedly throws the top card from his right hand before the bottom one.

Tornado Machine

What You Need

❑ Sandpaper
❑ Two clean, empty 2-liter soda bottles with caps
❑ Krazy Glue
❑ Black electrical tape
❑ Safety goggles
❑ Electric drill with ⅜-inch bit
❑ Water
❑ Blue food coloring
❑ Measuring spoons
❑ Silver glitter

What to Do

Sand the tops of the two bottle caps, and then glue the tops of the caps together with Krazy Glue. Let dry. Wrap black electrical tape around the circumference of the two caps to secure them together. With adult supervision and wearing safety goggles, drill a ⅜-inch hole through the center of the two caps.

Fill one of the bottles ⅔ full with water. Add five drops of blue food coloring and 1 tablespoon silver glitter. Thread the joined caps onto this bottle, and then thread the second bottle to the free end of the cap. Turn the Tornado Machine upside down so the blue glitter water pours into the empty bottle. Swirl the full bottle counterclockwise until a tornado funnel forms.

What Happens

The water whirls down into the empty bottle like a tornado.

Why It Works

The swirling water spins faster toward the hole, creating a vortex as the water molecules come closer to the center. The resulting outward force pushes the liquid out of the center, creating a funnel.

Bizarre Facts

■ According to a study of 304 tornadoes in the United States between 1950 and 1991, tornadoes are more likely to strike on May 16 than any other day of the year.

■ The winds of a tornado tend to whirl counterclockwise north of the equator and clockwise south of the equator.

■ The average tornado lasts less than thirty minutes and travels about twenty miles at ten to 25 miles per hour.

■ The overwhelming majority of tornadoes—approximately seven hundred a year—occur in the central and southeast United States, better known as Tornado Alley.

■ When a tornado touches the surface of an ocean or lake, it becomes a waterspout, sucking water up inside the spinning wind. Waterspouts appear to rise up out of the water like a sea monster, possibly explaining the origin of those legends.

I'll Get You, My Pretty

The tornado in the 1939 movie *The Wizard of Oz* was actually a funnel made of muslin stiffened with wire. To bring the muslin tornado to life, a prop man was lowered inside the muslin tube to pull the wires in and out.

Twirling Lines

What You Need

- Compass with pencil
- White posterboard
- Scissors
- Large nail
- Ruler
- Heavy black marker
- Turntable

What To Do

Using a compass with a pencil, make a circle 12 inches in diameter on the piece of posterboard. Using the scissors, cut out the disk.

Use the nail to punch a hole in the center of the disk.

Use the ruler and the black marker to make a thick black line along the diameter of the circle. Make three more heavy black lines along the diameter of the circle as if marking the posterboard into eight equal pieces like a pizza.

Place the posterboard disk in the center of the turntable, and spin the turntable rapidly.

What Happens

As the posterboard disk spins, the straight lines appear curved.

Why It Works

The brain, unable to properly interpret the rotating straight lines, simplifies the complex stimulus and perceives the rotating lines as circles.

Bizarre Facts

■ The lyrics to the pop song "Spinning Wheels," performed by Blood, Sweat, and Tears, tells of the spinning wheels of a merry-go-round.

House of Illusions

A house painted white appears larger than a house painted a dark color.

■ If you stare at the tires of a car as the car moves forward, the wheels appear to spin backward.

■ A person wearing a suit with vertical stripes appears thinner than that same person wearing a suit with horizontal stripes does.

Underwater Blue Beam

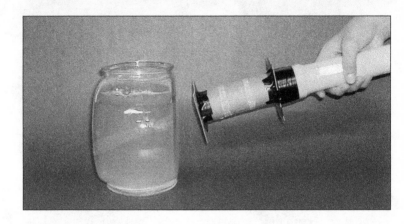

What You Need

- Sharpened pencil
- Two cardboard squares (3-by-3 inches)
- Scissors
- Black electrical tape
- Empty cardboard toilet paper tube
- Flashlight
- Fish bowl with two flat sides
- Measuring cup
- Water
- Milk
- Eyedropper
- Wooden spoon

What To Do

With the sharpened pencil, poke a hole in the center of both cardboard squares.

Using scissors, cut small strips of black electrical tape to adhere the two cardboard squares over the ends of the

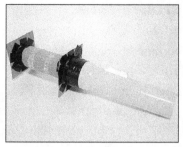

cardboard toilet paper tube. Then place one of the cardboard squares over the face of the flashlight and secure into place with strips of electrical tape. Wrap some electrical tape around the circumference of the head of the flashlight to prevent light from shining through the plastic or seeping through any cracks in the cardboard other than the punched hole.

Fill the fishbowl with water, add twenty drops of milk, and stir well with the wooden spoon. In a dark room, hold the lit flashlight six inches away from one of the flat sides of the fishbowl so the beam of light shines into the bowl. Look into the top of the bowl.

What Happens

A misty blue beam of light can be seen passing through the milky water. The beam of light emerging from the other side of the fishbowl appears to be orange.

How It Works

The beam of light is reflected and scattered in the water by the tiny colloidal particles of suspended fat from the milk, causing the beam of light to appear blue. This is called the Tyndall Effect, after British physicist John Tyndall. The light appears blue when the particles are smaller in diameter than $1/20$ the wavelength of light. The light emerging from the bowl appears orange because the blue light waves are scattered in the fishbowl.

221

Bizarre Facts

■ The Tyndall Effect is named after British scientist John Tyndall because he made the first comprehensive study of the effect, not because he discovered it. The effect was first witnessed in 1857 by English physicist Michael Faraday, best known for discovering the principle of electromagnetic induction.

■ In 1876, British physicist John Tyndall observed that a *Penicillium* mold slowed the growth of bacteria—fifty years before Sir Alexander Fleming's chemical work on penicillin.

■ You can also witness the Tyndall Effect when car headlights cut through the fog, when a flashlight beam is aimed through a room filled with dust or smoke, when sunlight comes in through a window and illuminates floats of dust, when beams of sunlight come through the clouds and illuminate the small drops of moisture in the air, or in a movie theater when the beam of light from the movie projector illuminates dust in the theater.

■ The Tyndall Effect can be seen in most episodes of the 1990s television series *The X-Files*, whenever secret agents Fox Moulder and Dana Scully investigate a dark place with flashlights.

■ The Tyndall Effect is used to determine whether a liquid is a suspension or a solution. The particles in a suspension are large enough to reflect or scatter light, making the light beam visible. A solution does not yield the Tyndall Effect.

■ In the 1600s, scientists claimed that light travels through an invisible, weightless, frictionless, stationary, omnipresent substance called ether that filled all space. Centuries later, in 1905, Albert Einstein published his theory of relativity, showing that light does not rely on the existence of ether.

Underwater Candle

What You Need

- Utility knife
- Corrugated cardboard box (roughly 1½-by-2 feet wide and 1½ feet high)
- Piece of corrugated cardboard (roughly 1½-by-2½ feet)
- Ruler
- Pencil
- Newspaper
- Paintbrush
- Flat black tempera paint
- Clear packaging tape
- Glass from a picture frame (about 8-by-10 inches)
- Candle
- Drinking glass
- Matches
- Pitcher
- Water

What to Do

With adult supervision, use the utility knife to carefully cut the flaps from the top of the box, and fit the long piece of cardboard snugly upright diagonally inside the box (from the front left-hand corner to the back right-hand corner), dividing the box equally into two triangular halves.

Carefully measure and cut a small viewing window (6-inches square) in the front left half of the box where the dividing cardboard meets the corner.

Remove the divider from the box and cut a window about 7-by-9 inches that will line up with the window on the box, allowing you to look straight through to the left rear side of the box. Spread newspaper on the tabletop and paint the divider and the inside of the box with the flat black tempera paint. When the paint is dry, use the packaging tape to secure the 8-by-10-inch glass over the opening in the divider. Place the divider back in the box.

Mount the candle in the center of the first compartment. Place the drinking glass in the center of the rear compartment.

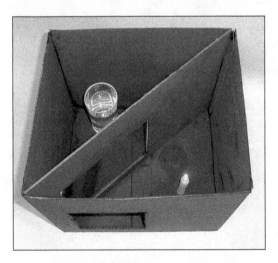

Light the candle. Look through the viewing window and through the glass window. Align the candle and the glass until the candle appears in the center of the glass.

Have a friend look through the viewing window. Slowly pour the pitcher of water into the empty glass in the rear compartment.

What Happens

Your friend sees a candle burning in a glass of water.

Why It Works

The light reflecting from the front of the glass window reflects the image of the candle in the glass. This optical illusion is created because you can still see through the window to the glass with the water.

Bizarre Facts

■ This optical illusion is used in circus sideshows to create the illusion of a human turning into a skeleton or monster.

■ In 1862, after fifteen years of work, French physicist Jean Foucault used a candle and two mirrors to measure the speed of light at 187,000 miles per second. Modern methods have refined the figure to 186,282 miles per second.

■ Our eyes remain the same size from birth, but our nose and ears never stop growing.

It's a Bird, It's a Plane

In comic books, when Superman uses his X-ray vision to see through concrete walls, X-rays shoot from his eyes. Living creatures do not see by radiating light from their eyes. (X-rays are a more energetic form of light.) Instead, light enters the eye, is refracted by the cornea, passes through the pupil, is refracted by the lens, and forms an image on the retina, which is rich in nerve cells that are stimulated by light. To see through concrete walls, the nerve cells in Superman's retinas would have to be stimulated by X-rays.

Underwater Fireworks

What You Need

❑ Large, clear glass bowl
❑ Water
❑ 1 tablespoon vegetable oil
❑ Paper cup

❑ Food coloring (red, blue, and green)
❑ Spoon

What to Do

Fill the bowl with water. Pour the cooking oil into the paper cup. Add four drops of each food coloring color. Mix the oil and colors thoroughly with the spoon. Pour the colored oil mixture into the water in the bowl. Observe for ten minutes.

What Happens

Small pools of oil spotted with tiny spheres of color float to the surface of the water, exploding outward and creating flat circles of color on the surface of the water. Long streamers of color then sink down through the water, like a fireworks display.

Why It Works

Oil and water are *immiscible*—meaning they do not mix, but separate into layers because of the different polarity of their molecules. The oil rises to the surface because it is less dense than the water. Since the water-based food coloring does not dissolve in oil, it remains in tiny spheres throughout the oil on the water's surface and then sinks through the oil layer and dissolves in the water below, creating long streamers of color.

Bizarre Facts

■ In the tenth century, the Chinese discovered that three common kitchen ingredients—saltpeter, sulfur, and charcoal—were explosive when combined. When packed into a bamboo tube and ignited, the mixture rocketed skyward and exploded, lighting up the sky.

■ In fireworks, sodium compounds produce yellow light, strontium and lithium salts emit red light, copper gives blue light, and barium creates green light.

■ In 1988, the world's longest fireworks display—measuring 18,777 feet long, consisting of 3,338,777 firecrackers and 1,468 pounds of gunpowder—was ignited in Johor, Malaysia, and burned for nine hours and 27 minutes.

■ On July 15, 1988, the world's largest firework—weighing 1,543 pounds and measuring 1,354.7 inches in diameter—was exploded over Hokkaido, Japan, bursting to a diameter of 3,937 feet.

Bye Bye Böögg

On the third Monday in April, the citizens of Zurich, Switzerland, stuff the Böögg, an enormous cotton-wool snowman, with fireworks, mount it to a pole amid a pyre of brushwood, and ignite the bonfire. Riders on horses circle the fire until the fireworks blow the Böögg to bits.

Upside-Down Water

What You Need

❑ Drinking glass
❑ Water

❑ Cardboard (4-by-4 inches)

What To Do

Fill the drinking glass to the brim with water. Gently place the piece of cardboard over the mouth of the drinking glass. Holding the cardboard in place, turn the drinking glass upside down. Release the piece of cardboard.

What Happens

The cardboard sticks to the rim of the glass and holds the water in the glass.

Why It Works

Atmospheric pressure—the force exerted by the weight of the air—holds the cardboard in place.

Bizarre Facts

■ Atmospheric pressure is the force exerted by the weight of air molecules. At sea level, the earth's atmosphere presses against you with a force of 14.7 pounds per square inch. The force exerted on one square foot is more than one ton. But we're not squished by this because the molecules of our own bodies exert an equal and opposite force.

■ Your ears pop and you need to breathe more rapidly on top of a tall mountain because the atmospheric pressure is less than at sea level. In other words, there are less air molecules (causing you to breath faster to fill your lungs with oxygen). At high altitudes, the air molecules also weigh less (causing your ears to pop to balance the pressure between the outside and inside of your ears).

■ At sea level the density of air is approximately eight ounces per square foot. Both pressure and density decrease by about a factor of ten for every ten-mile increase in altitude.

■ The Channel Tunnel, nicknamed the Chunnel, between Cheriton, England, and Fréthum, France, is 31 miles long, with 24 of those miles underwater. The Seiken Rail Tunnel between the Japanese islands of Honshu and Hokkaido is 33.5 miles long, with 14.5 of those miles underwater, making it the longest railroad tunnel in the world. The Chunnel, however, is the longest underwater railroad tunnel in the world.

Volcano Madness

Volcano at the Beach

What You Need

- Clean, empty 1-liter soda bottle
- 1 cup water
- 1 tablespoon baking soda
- 1 tablespoon liquid dish-washing detergent
- Red food coloring
- Plastic dinosaurs or army soldiers
- 1 cup white vinegar

What to Do

Fill the bottle with the water, baking soda, liquid dishwashing detergent, and ten drops of red food coloring. Place the bottle on

the beach and build a sand volcano around the bottle. Place the plastic dinosaurs or soldiers around the volcano. Pour the vinegar into the bottle and scream, "Volcano! Volcano! Volcano!"

What Happens

Red foam will bubble and spray out the top of the volcano and down the mountain of sand, covering the dinosaurs or soldiers.

Why It Works

The baking soda (a base) reacts with the vinegar (an acid) to produce carbon dioxide, generating foam and forcing the liquid out of the bottle.

The Volcano Farm

On February 20, 1943, a volcano began forming from a crack in the earth in a farmer's cornfield in Paricutín, Mexico (near the southwestern city of Uruapan). Within six days, the volcanic material formed a cinder cone over five hundred feet high. Two months later, the cone reached one thousand feet. The lava destroyed the village of Paricutín and San Juan Parangaricutiru. Today the volcano, which ceased activity in 1952, stands 1,345 feet high.

Exploding Volcano

What You Need

- ❏ Scissors
- ❏ Metal screening
- ❏ Needle
- ❏ Thread
- ❏ Newspaper
- ❏ Circle of ¼-inch thick
    pinewood, 1 foot in
    diameter
- ❏ Hammer
- ❏ Carpet tacks (or staple gun
    and staples)
- ❏ Clean, empty, 3-ounce cat
    food can
- ❏ Measuring cup
- ❏ Plaster of Paris
- ❏ Water
- ❏ Bowl
- ❏ Paintbrush
- ❏ Brown paint
- ❏ Shellac
- ❏ Safety goggles
- ❏ Rubber gloves
- ❏ Fire extinguisher
- ❏ Ammonium dichromate
    (from a chemical supply
    store)
- ❏ Matches

What to Do

With adult supervision and using scissors, form a cone with the
metal screening, and sew it together with the needle and thread.
Fill the cone with crumpled-up newspaper, and tack it down to
the pine board. Set the cat food can in the cone of the volcano.

Cover the tabletop with newspaper. Cut newspaper strips 1 to 2 inches in width. Mix plaster of Paris and water in a bowl, according to the package instructions. Dip each newspaper strip into the plaster, gently pull it between your fingers to remove excess plaster, and apply it to the screening until it is completely covered. Let it dry, and then paint it with the brown paint. Spray with shellac to seal the volcano. Let dry thoroughly.

Set the volcano outdoors. Crumple up a small piece of newspaper, put it in the cat food can, and wearing safety goggles and

rubber gloves, pour some ammonium dichromate over it. With an adult watching and a fire extinguisher on hand, light the newspaper with a match, and quickly back away. (Instead of ammonium dichromate, you can place a smoke bomb from page 196 in the volcano and ignite it.)

What Happens

The volcano flares up, sparks, sputters, and sends fluffy green lava spilling over the sides.

233

Miracle on Bali

Besakih Temple, the holiest temple on the island of Bali, sits at the foot of the Gunung Agung volcano. On this holy spot, the *Eka Dasa Rudra* festival is held once every one hundred years. On the last day of the celebration in 1963, for the first time in centuries, the volcano erupted, killing thousands of people and sending lava coursing over eastern Bali. Miraculously, Besakih Temple remained untouched.

Why It Works

When ignited, ammonium dichromate (an orange, crystalline solid) spews sparks, a fluffy green solid (chromium oxide), steam (water vapor), and heat (nitrogen gas).

Underwater Volcano

What You Need

- Large pot
- Cold water
- Ice cubes
- Spoon
- Clean, empty baby food jar
- Hot tap water
- Red food coloring
- Marbles

What to Do

Fill the pot with cold water and put in some ice cubes. When the water is sufficiently cold, fish out the ice cubes with the spoon. Fill the baby food jar ¾ full with hot tap water and add five drops of red food coloring. Stir well. Add five or six marbles to give the jar some weight. Place the bottle in the bottom of the pot.

What Happens
The red water slowly "erupts" from the jar toward the surface of the water.

Why It Works
Heat rises. Hot water is lighter than cold water because the molecules in hot water are farther apart than those in cold water.

Bizarre Facts
■ Pumice, a natural glass that comes from lava, is widely used to remove calluses.

■ In Reykjavik, Iceland, most people heat their homes with water piped from volcanic hot springs.

■ When Mount Tambora in Indonesia erupted in 1815, it released six million times more energy than an atomic bomb and killed over 12,000 people.

■ In 1963, an underwater eruption began forming the island of Surtsey in the North Atlantic Ocean. After the last eruption of lava in 1967, the island covered more than one square mile.

■ History's most famous volcanic eruption, Mount Vesuvius, took place in 79 C.E. and destroyed the Italian towns of Herculaneum, Pompeii, and Stabiae. Pompeii remained untouched beneath the ash deposits for almost 1,700 years.

■ The word *volcano* comes from Vulcan, the Roman god of fire who lived beneath a volcanic island called Vulcano near Italy.

■ The *Voyager* mission's photographs of Io, Jupiter's third-largest moon, showed ten active volcanoes, some of them erupting—making Io more volcanically active than earth.

Wax Snowflakes

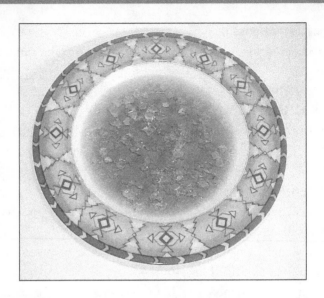

What You Need

- Bowl
- Cold water
- Blue food coloring
- Spoon
- Matches
- Candle

What To Do

Fill the bowl with cold water, add five drops of blue food coloring, and stir well. With adult supervision, light the candle and hold it two feet above the bowl of water, allowing hot wax to drip into the water.

What Happens

The drops of wax float on the surface of the water and look like snowflakes.

How It Works

The moment the liquid hot wax splatters on the surface of the cold water, the wax immediately solidifies, suggesting the appearance of snowflakes.

Bizarre Facts

■ When the Coca-Cola Company introduced Coke in China, the name was transliterated as "Kekoukela," without any regard for the actual meaning of the sounds in Chinese. In different dialects, the phrase "Kekoukela" translates as "bite the wax tadpole" or "female horse stuffed with wax."

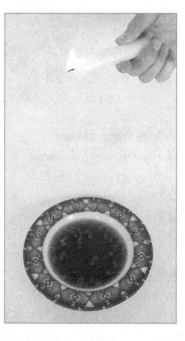

■ Only 1 percent of all snowflakes are symmetrical.

■ In 1611, German mathematician and astronomer Johannes Kepler wrote a pamphlet entitled *On the Six-Cornered Snowflake*, but was unable to explain the hexagonal shape of snowflakes.

■ The world record for most snowfall in 24 hours was made on February 7, 1963 when 78 inches of snow fell at Mile 47 Camp, Cooper River Division 4, Alaska.

X-ray Glasses

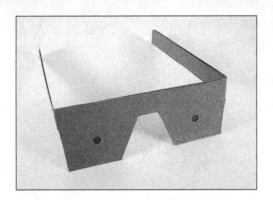

What You Need

- Two strips of red poster-board (2-by-17 inches)
- Ruler
- Pencil
- Scissors
- Hole puncher
- Black feather
- Elmer's Glue-All
- Two hardcover books
- Bright light

What To Do

Fold the first strip of red posterboard in half so the 17-inch length is now 8½ inches long. Place the folded strip on a tabletop with the fold facing to your right. From the top edge of the posterboard strip, use the ruler to measure down ⅝ inch and draw a line across the width of the poster-board strip. From the right-hand fold, measure in ⅞ inch and draw a vertical line across the height of the posterboard strip. From

 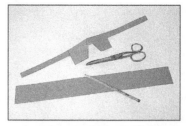

the bottom of this line, draw a diagonal line to the top right-hand corner. From the right-hand fold, measure in 2½ inches and draw a vertical line across the height of the posterboard strip.

Using the scissors, cut out the large rectangle in the bottom left-hand corner, cutting through both layers of posterboard. Cut out the quadrangle in the bottom right-hand corner, cutting through both layers of posterboard.

Open the pair of glasses and use it as a template on the second strip of red posterboard to create an identical pair of glasses.

Hold the two pairs of red-posterboard glasses together so they line up perfectly, and using the hole puncher, punch a hole in the center of each lens.

Cut two pieces from the feather, each one inch long. Glue the first piece of feather to one of the red-posterboard glasses so that the feather covers one of the holes (without getting any glue on the part of the feather that covers the hole).

Glue the second piece of feather over the second hole of that same pair of glasses.

Glue the second pair of red-posterboard glasses to the first pair so the feathers are sandwiched between them and the holes in the lenses line up. Place the two hardcover books on top of the glasses to press the two pieces together until the glue dries.

When the glue dries, fold back the arms of the glasses, wear the X-ray glasses, and under a bright light, look at your hand through the glasses.

What Happens

You see what appear to be the bones in your fingers.

How It Works

The eye sees different wavelengths of light as different colors. Light travels in waves, which usually travel in a straight line. However, when light waves pass through a slit, they diffract (spread out) into curving waves. When light waves pass through several narrow slits (like the numerous spaces created by the complex branching pattern of the feather), they interfere with each other. Where the crest (peak) of one wave meets the crest of another wave—or where the trough (low point) of one wave meets the trough of another wave—the two waves combine and form a bright spot of light. Where a crest meets a trough, the two waves cancel each other out, leaving a dark spot.

Some of the light which would normally be stopped by the edges of your fingers gets bent and reaches your eye, making the edges of your fingers appear semi-transparent. Meanwhile, light from the center of the fingers is not bent enough to reach your eyes, so the center of your fingers appear solid, resembling the bones of your fingers.

Bizarre Facts

■ X-ray Specs were commonly advertised by the Johnson-Smith Company in the back of comic books.

■ Birds shed their feathers and grow a new set at least once a year.

■ Feathers are made of *keratin*, a substance also found in the hair of mammals and the scales of fish and reptiles.

■ In 1666, English scientist Sir Isaac Newton theorized that light consists of tiny particles called corpuscles that travel in straight lines through space. In 1801, English physicist Thomas Young proved Newton wrong, demonstrating that particles of light travel in waves.

■ The 1963 Roger Corman horror film *X—The Man with X-ray Eyes* stars Ray Milland as a scientist named Xavier who gives himself X-ray vision and goes insane. The movie co-stars comedian Don Rickles.

Dumb Luck

In 1895, German physics professor William Konrad Roentgen discovered X-rays by accident when he noticed that cathode rays caused a sheet of paper coated with barium platinocyanide to glow—even when the sheet of paper was taken into the next room. Roentgen called the rays X-rays because of their mysterious nature.

Yogurt Factory

What You Need

- 1 quart of half-and-half
- Glass mixing bowl
- Cooking thermometer
- Spoon
- 3 tablespoons of Dannon
  Nonfat plain yogurt
- Clean, empty 1-quart glass
  jar with lid
- Insulated picnic cooler

What To Do

Pour the half-and-half into the glass mixing bowl.

With adult supervision and using a cooking thermometer, heat the half-and-half in a microwave oven at 50 percent power for one minute at a time until the temperature of the half-and-half reaches 180 degrees Fahrenheit. (Use the spoon to stir the liquid between time breaks to prevent scalding and skim off any film from the surface.) Remove from the microwave and let cool to about 115 degrees Fahrenheit.

Add three tablespoons of yogurt, mix well, pour into the jar, and seal the lid tightly. Place the warm jar inside the insulated picnic cooler, close the lid, and let sit undisturbed for eight hours. Refrigerate when ready.

What Happens
You've made plain yogurt. If you wish, add vanilla extract, or strawberries, peaches, or raspberries to taste.

How It Works
Heating the half-and-half kills any bacteria that might otherwise compete with the yogurt cultures. Yogurt contains living bacteria called *Lactobacillus acidophilus*, which multiply exponentially in warm milk. Three tablespoons of yogurt from this batch can be used to start a new batch, ideally within five days.

Bizarre Facts
■ Adding the three tablespoons of yogurt to the milk when it is above 155 degrees Fahrenheit may kill the yogurt cultures, which would then prevent the yogurt from forming.

■ Adding more than three tablespoons of yogurt to the warm milk will cause overcrowded bacillus, resulting in sour, watery yogurt.

Crying Over Spilled Milk

For their satirical operetta *H.M.S. Pinafore* (1878), W. S. Gilbert and Arthur Sullivan created the pithy lyric, "Things are seldom what they seem; skim milk masquerades as cream."

Bibliography

- *Bigger Secrets* by William Poundstone (New York: Houghton Mifflin, 1986)
- *Biggest Secrets* by William Poundstone (New York: Quill, 1993)
- "Bizarre Stuff You Can Make in Your Kitchen" by Brian Carusella, http://bizarrelabs.com
- *The Book of Lists* by David Wallechinsky, Irving Wallace, and Amy Wallace (New York: Bantam, 1977)
- *Can It Really Rain Frogs?* by Spencer Christian (New York: John Wiley & Sons, 1997)
- *Chariots of the Gods?* by Erich Von Däniken (New York: Bantam, 1970)
- *The Concise Oxford Dictionary of Proverbs* by John Simpson (Oxford, England: Oxford University Press, 1992)
- *Dictionary of Trade Name Origins* by Adrian Room (London: Routledge & Kegan Paul, 1982)
- *The Dorling Kindersley Science Encyclopedia* (New York: Dorling Kindersley, 1994)
- *Duct Tape Book Two—Real Stories* by Jim and Tim (Duluth, Minnesota: Pfeifer-Hamilton, 1995)
- *Einstein's Science Parties* by Shar Levine and Allison Grafton (New York: John Wiley and Sons, 1994)
- *Elements of Psychology: A Briefer Course* by David Krech, Richard S. Crutchfield, and Norman Livson (New York: Knopf, 1970)
- *Eyewitness Books: Crystal and Gem* by R. F. Symes and R. R. Harding (New York: Knopf, 1991)
- *Famous American Trademarks* by Arnold B. Barach (Washington, D.C.: Public Affairs Press, 1971)

■ *The Guinness Book of Records* edited by Peter Matthews (New York: Bantam Books, 1998)

■ "Have a Problem? Chances Are Vinegar Can Help Solve It" by Caleb Solomon, *Wall Street Journal*, September 30, 1992

■ *How in the World?* by the editors of *Reader's Digest* (Pleasantville, New York: Reader's Digest, 1990)

■ *How to Play with Your Food* by Penn Jillette and Teller (New York: Villard, 1992)

■ *How to Spit Nickels* by Jack Mingo (New York: Contemporary Books, 1993)

■ *Janice VanCleave's 200 Gooey, Slippery, Slimy, Weird & Fun Experiments* by Janice VanCleave (New York: John Wiley & Sons, 1993)

■ *The Joy of Cooking* by Irma S. Rombauer and Marion Rombauer Becker (New York: Bobbs-Merrill, 1975)

■ *Jr. Boom Academy* by B. K. Hixson and M. S. Kralik (Salt Lake City, Utah: Wild Goose, 1992)

■ *Marbling* by Diane Vogel Maurer with Paul Maurer (New York: Friedman Fairfax, 1994)

■ *Marbling Techniques* by Wendy Addison Medeiros (New York: Watson Guptill, 1994)

■ *Martin Gardner's Science Tricks* by Martin Gardner (New York: Sterling, 1998)

■ *Modern Chemical Magic* by John D. Lippy, Jr., and Edward L. Palder (Harrisburg, Pennsylvania: Stackpole, 1959)

■ *More Science for You: 112 Illustrated Experiments* by Bob Brown (Blue Ridge Summit, Pennsylvania: TAB Books, 1988)

■ *100 Make-It-Yourself Science Fair Projects* by Glen Vecchione (New York: Sterling, 1995)

■ *PADI Open Water Diver Manual* (Rancho Santa Margarita, California: International PADI, 1999)

■ *Panati's Extraordinary Origins of Everyday Things* by Charles

Panati (New York: Harper & Row, 1987)

■ *Paper Art* by Diane Maurer-Mathison with Jennifer Philippoff (New York: Watson-Guptill, 1997)

■ *Paper Craft School* by Clive Stevens (Pleasantville, New York: Reader's Digest, 1996)

■ *Physics for Kids* by Robert W. Wood (Blue Ridge Summit, Pennsylvania: Tab Books, 1990)

■ *Reader's Digest Book of Facts*, edited by Edmund H. Harvey, Jr., (Pleasantville, New York: Reader's Digest, 1987)

■ *Reader's Digest How Science Works* by Judith Hann (Pleasantville, New York: Reader's Digest, 1991)

■ *Ripley's Believe It or Not! Encyclopedia of the Bizarre* by Julie Mooney and the editors of *Ripley's Believe It or Not!* (New York: Black Dog & Leventhal, 2002)

■ *The Safe Shopper's Bible* by David Steinman and Samuel S. Epstein, M.D. (New York: Macmillan, 1995)

■ *Science Fair Survival Techniques* (Salt Lake City, Utah: Wild Goose, 1997)

■ *Science for Fun Experiments* by Gary Gibson (Brookfield, Connecticut: Copper Beech Books, 1996)

■ *Science Projects about Light* by Robert Gardner (Hillside, New Jersey: Enslow, 1994)

■ *Science Wizardry for Kids* by Margaret Kenda and Phyllis S. Williams (Hauppauge, New York: Barron's, 1992)

■ *Shake, Rattle and Roll* by Spencer Christian (New York: John Wiley & Sons, 1997)

■ *Shout!—The Beatles in Their Generation* by Philip Norman (New York: Warner Books, 1981)

■ *South-East Asia Handbook* by Stefan Loose and Renate Ramb (Berlin, Germany: Stefan Loose Travel Books, 1983)

■ *Structure of Matter* by the editors of *Time Life* (New York: Time Life, 1992)

■ *Sure-to-Win Science Fair Projects* by Joe Rhatigan with Heather Smith (New York: Lark Books, 2001)

■ *333 Science Tricks & Experiments* by Robert J. Brown (Blue Ridge Summit, Pennsylvania: TAB Books, 1984)

■ *365 Simple Science Experiments* by E. Richard Churchill, Louis V. Loeschnig, and Muriel Mandell (New York: Black Dog & Leventhal, 1997)

■ *200 Illustrated Science Experiments for Children* by Robert J. Brown (Blue Ridge Summit, Pennsylvania: TAB Books, 1987)

■ *The Ultimate Duct Tape Book* by Jim and Tim (Duluth: Minnesota: Pfeifer-Hamilton, 1998)

■ *The Way Science Works* by Robin Kerrod and Dr. Sharon Ann Holgate (New York: DK Publishing, 2002)

■ *Weird Europe: A Guide to Bizarre, Macabre, and Just Plain Weird Sights* by Kristan Lawson and Anneli Rufus (New York: St. Martin's Press, 1999)

■ *What Makes the Grand Canyon Grand?* by Spencer Christian (New York: John Wiley & Sons, 1998)

■ *Why Did They Name It . . . ?* by Hannah Campbell (New York: Fleet, 1964)

Acknowledgments

I am grateful to Jennifer Repo, Michelle Howry, Erin Stryker, Dolores McMullen, and John Duff at Perigee Books for their enthusiasm, passion, and excitement for publishing my three Mad Scientist Handbooks that help compose this compilation. I am also indebted to Amy Schneider, Cari Luna, and Sheila Moody for their astounding copyediting expertise.

A very special thanks to Jeremy Solomon at First Books for sharing my unbridled exuberance for those original books and for his inspiring professionalism and zealous perseverance.

I am also grateful to my father, Bob Green, for helping me build a volcano in the fourth grade; Debbie Green for her research and proofreading expertise; Carrie Bruder for the beautiful photograph on the front cover; Tony Salzman for sharing his first-hand experiences with steel wool and 9-volt batteries; Howard Gershen for the paper cup boiler; Jeremy Wolff for perfecting the hydrogen balloon experiment; Mark, Debra, Marissa, and Jason Jaffe for providing the inspiration for Mayonnaise Madness; Jim Parish for suggesting the idea in the first place; Girl Scout Troop 1380 for helping me perfect some of the experiments; and Katie Loggia, Emily Graham, Alexa Cohen, Rebecca Leon, and Jay Bruder for their able assistance.

Above all, all my love to my wife, Debbie, and my daughters, Ashley and Julia, for eagerly assisting me with the experiments in this book in our garage and for wisely refusing to help me clean up the resulting mess.

About the Author

Joey Green got Barbara Walters to make green slime on *The View*, Jay Leno to shave with Jif peanut butter on *The Tonight Show*, Rosie O'Donnell to mousse her hair with Jell-O on *The Rosie O'Donnell Show*, and Katie Couric to drop her diamond engagement ring in a glass of Efferdent on *Today*. He has been seen polishing furniture with Spam on *Dateline NBC*, cleaning a toilet with Coca-Cola in the *New York Times*, and washing his hair with Reddi-wip in *People*. Green, a former contributing editor to *National Lampoon* and a former advertising copywriter at J. Walter Thompson, is the author of more than fifty books, including *Contrary to Popular Belief*, *Sarah Palin's Secret Diary*, and *Selling Out: If Famous Authors Wrote Advertising*.

A native of Miami, Florida, and a graduate of Cornell University, he wrote television commercials for Burger King and Walt Disney World, and won a Clio Award for a print ad he created for Eastman Kodak. He backpacked around the world for two years on his honeymoon, and lives in Los Angeles with his wife, Debbie, and their two daughters, Ashley and Julia.

> **Visit Joey Green on the internet at**
> **www.wackyuses.com**

CPSIA information can be obtained
at www.ICGtesting.com
Printed in the USA
BVHW071135170920
588935BV00003B/245

9 780977 259069